Current Topics in Environmental Health and Preventive Medicine

Series Editor

Takemi Otsuki
Kawasaki Medical School
Matsushima, Kurashiki-shi
Okayama, Japan

Current Topics in Environmental Health and Preventive Medicine, published in partnership with the Japanese Society of Hygiene, is designed to deliver well written volumes authored by experts from around the globe, covering the prevention and environmental health related to medical, biological, molecular biological, genetic, physical, psychosocial, chemical, and other environmental factors. The series will be a valuable resource to both new and established researchers, as well as students who are seeking comprehensive information on environmental health and health promotion.

More information about this series at http://www.springer.com/series/13556

Seiichiro Himeno · Keiko Aoshima
Editors

Cadmium Toxicity

New Aspects in Human Disease, Rice Contamination, and Cytotoxicity

Springer

Editors
Seiichiro Himeno
Faculty of Pharmaceutical Sciences
Tokushima Bunri University
Tokushima
Japan

Keiko Aoshima
Hagino Hospital
Toyama
Japan

ISSN 2364-8333 ISSN 2364-8341 (electronic)
Current Topics in Environmental Health and Preventive Medicine
ISBN 978-981-13-3629-4 ISBN 978-981-13-3630-0 (eBook)
https://doi.org/10.1007/978-981-13-3630-0

Library of Congress Control Number: 2019930140

This Springer imprint is published by the registered company Springer Nature Singapore Pte Ltd.
The registered company address is: 152 Beach Road, #21-01/04 Gateway East, Singapore 189721, Singapore

Preface

Cadmium (Cd) is a well-known global environmental pollutant. In Japan, however, the problem of environmental pollution by Cd has appeared as severe cases of human poisoning, known as *itai-itai* disease, which occurred in a confined local area. The cause of *itai-itai* disease has been linked to the Cd-contaminated mine wastewater leaked from the Kamioka zinc mine, which is located upstream of the Jinzu River in Toyama Prefecture. On a worldwide scale, the severity of the symptoms of *itai-itai* disease has created a sense of urgency, leading to a global recognition of Cd as a serious environmental pollutant. Until now, however, the details of the clinical, epidemiological, and pathological aspects of *itai-itai* disease have not been sufficiently recognized even among the scientists involved in metal toxicology.

As a volcanic country, Japan is blessed with many mining resources. Because many mines and smelting factories had been operating in the mountainous areas in the past, the water of the river, in addition to the Jinzu River, used for the irrigation of paddy fields had been polluted by Cd. Although strict restriction and management by the government during the last decades have improved the situation, the paddy fields still contain low-level Cd in some areas. Since the rice is a staple food among Japanese, the consumption of rice contributes significantly to total Cd intake among Japanese compared with other countries. The National Institute of Health Sciences has reported (2015) that rice is the major source of Cd consumption among Japanese, accounting for about 30% (previously more than 40%) of the total Cd consumption from diet (Chap. 1). Thus, the reduction of Cd in rice is still an important issue in agricultural policy in Japan. In recent years, advances in soil sciences and plant physiology have led to a number of successful countermeasures against Cd accumulation in rice in Japan.

The contributors to this volume were chosen for their expertise on various aspects of Cd toxicity in Japan. Thus the contents include not only medical and biochemical considerations of Cd toxicity but also agricultural science investigations into the mitigation of Cd accumulation in rice.

Part I (Overview; Chap. 1) provides a historical overview of *itai-itai* disease, including the social aspects, and discusses the roles of regulatory science in establishing government policy at a time when causal relationships between the environmental pollutant and the disease are not sufficiently proven. The current exposure levels of Cd in Japan are also given in this part.

Part II (Health Effects of Cadmium in Human Populations and its Relation to Cadmium Exposure Levels; Chaps. 2–7) deals with the clinical and epidemiological aspects of Cd poisoning in Japan, including *itai-itai* disease. In the case of *itai-itai* disease, recent findings on the clinical, biochemical, pathological, and epidemiological aspects of the disease obtained by long-term follow-up studies are presented. Epidemiological studies in areas other than Toyama Prefecture have examined the morbidity and mortality patterns of the residents in Cd-polluted areas, and a study in Akita Prefecture demonstrated the renal dysfunctions among aged populations consuming Cd-polluted rice. Surveys of general populations have revealed the time-dependent decreases in dietary intake of Cd.

Part III (Rice Contamination by Cadmium and its Remediation; Chaps. 8–11) is a unique part of this book. The reports presented in this part provide scientific evidence regarding the transport mechanisms of Cd from the soil to rice roots, the mitigation methods for reducing rice Cd accumulation based on soil science, and a genetic approach for producing mutated rice that does not accumulate Cd. The trade-off problem of Cd and inorganic arsenic accumulation into rice is also discussed here.

Part IV (New Aspects in the Mechanisms of Cadmium Toxicity; Chaps. 12–14) includes mechanistic studies on the cytotoxicity and transport of Cd. Previously, many studies were focused on the roles of metallothionein in the toxicity and transport of Cd. In this part, however, newly discovered roles of ubiquitin-proteasome pathways in Cd-induced apoptosis, the involvement of multiple metal transporters in cellular transport of Cd, and the effects of Cd on the regulation of phosphorus in the kidney are described.

Thus, starting from a historical overview of *itai-itai* disease, this book covers multiple phases of Cd toxicity, including the clinical and epidemiological aspects of human Cd poisoning, mechanisms of Cd accumulation in rice and its mitigation by novel technology, and biochemical mechanisms of Cd toxicity and transport. *Itai-itai* disease is a local problem in Japan caused by local Cd pollution, but we are confident that the comprehensive and cutting-edge information on the medical, agricultural, and biochemical aspects of Cd toxicity shown in this book will be useful for all readers interested in the solution of local and global Cd pollution. Finally, we are grateful to all the participating authors for their excellent papers and also to Springer and The Japanese Society for Hygiene for giving us the opportunity to publish this book.

Tokushima, Japan Seiichiro Himeno
Toyama, Japan Keiko Aoshima

Contents

Part I
Overview

Chapter 1
Historical Lessons on Cadmium Environmental Pollution Problems in Japan and Current Cadmium Exposure Situation

Keiko Aoshima and Hyogo Horiguchi

Abstract In 1968, the Ministry of Health and Welfare acknowledged that *itai-itai* disease, which occurred frequently in the Jinzu River basin of Toyama Prefecture, was a pollution disease. This chapter explores three topics related to *itai-itai* disease that have not been examined before with the aim of extracting historical lessons for the cadmium (Cd) environmental pollution problem in Japan. The first topic deals with the progress made by Dr. Kinichi Yoshioka, who first reported Cd environmental contamination, and suggested that *itai-itai* disease could be caused by chronic Cd poisoning. This topic focuses on the significance of ecological and epidemiological studies concerned with the actual circumstances surrounding heavy metal pollution (mainly Cd), in particular, the exposure route to the local residents. The second topic introduces the decision made by the administrative official that summarizes the opinion of the Ministry of Health and Welfare that "*itai-itai* disease is a pollution disorder caused by chronic Cd poisoning." This opinion became the starting point for government pollution countermeasures as well as its policy decision-making process and illustrates that Cd environmental pollution presented a critical problem to regulatory science. The third topic analyzes "Cadmium Studies in Japan: A Review" published in 1978 and "Interim Report Summary" published in 1989. The progress made and the challenges encountered by the "*Itai-itai* Disease and Chronic Cd Poisoning Study Group" are then summarized. Since 1969, this group has been commissioned by the Ministry of Health and Welfare or the Ministry of the Environment to conduct studies to determine the relation between *itai-itai* disease and Cd poisoning. In the present circumstances surrounding Cd exposure in Japan,

K. Aoshima (✉)
Hagino Hospital, Toyama, Toyama, Japan
e-mail: keiaoshima@po4.nsk.ne.jp

H. Horiguchi
Department of Hygiene, Kitasato University School of Medicine,
Sagamihara, Kanagawa, Japan

S. Himeno, K. Aoshima (eds.), *Cadmium Toxicity*, Current Topics in Environmental Health and Preventive Medicine,
https://doi.org/10.1007/978-981-13-3630-0_1

the daily Cd intake from the diet of general residents has decreased from 31.1 μg in 1981 to 17.8 μg in 2015 (a reduction of over 30% in over 30 years). The reason for this was considered to be a decrease in the intake of rice, a staple Japanese food, and a reduction in the Cd concentration in rice due to improvements in rice cultivation methods.

Keywords Cadmium · Environmental pollution · *Itai-itai* disease · Regulatory science · Daily intake in Japan

1.1 Historical Lessons from the Cadmium Environmental Pollution Problems in Japan

1.1.1 Introduction

Itai-itai disease is the most severe type of chronic cadmium (Cd) intoxication from environmental pollution generated in the Jinzu River basin, Toyama Prefecture. Its primary pathological condition is proximal tubular osteomalacia [1]. In chronic Cd intoxication, the early onset of proximal tubular dysfunction occurs, and the urinary excretion level of glucose, amino acids, phosphorus, uric acid, and low molecular weight proteins such as β_2-microglobulin increases [2]. Multiple proximal tubular dysfunction ("Cd nephropathy") is frequently experienced by the residents in Cd-contaminated areas of the Jinzu River basin [3]. However, the national government (Ministry of the Environment) only recognizes *itai-itai* disease as a pollution disease and not Cd nephropathy.

Along the Takahara River in the upper reaches of the Jinzu River lies Kamioka Mine, one of Japan's leading lead-zinc mine smelters ("Mitsui Mining & Smelting Co., Ltd."), which is currently being operated by Kamioka Mining & Smelting Co., Ltd. Due to the slag discarded from the mines, the rice crops in the lower reaches of the Jinzu River have suffered damage since the end of the Meiji era (around 1910); farmers have frequently protested against the miners and government administrators, requesting that countermeasures be taken to prevent further damage. In 1919, the Toyama Prefecture Kaminiikawa County Agricultural Association requested the Ministry of Agriculture and Commerce's Nishigahara Agricultural Experiment Station to analyze the concentration of zinc in the irrigated soil. The results of the analysis along with a proposal for Kamioka Mine to remove mineral poisons were reported to the Minister of Agriculture and Commerce and to the Governor of Toyama Prefecture.

Itai-itai disease has been occurring since the Taisho era (1912–1925), and it frequently occurred throughout the Showa era (1935–1944). However, the social circumstances during wartime implied that "mine pollution" was a taboo; therefore, no scientific elucidation or governmental measures were undertaken. Finally, in 1946, after the end of the Second World War, local residents asked Kanazawa University to investigate "a strange disease occurring frequently." However, more than 10 years were required for this disease to become a socially recognized problem, and the term "*itai-itai* disease" finally came into existence in 1955. Early studies focused on clinical and medicinal research, including clinical findings, pathology, diagnosis, and treatment. Factors, such as age, malnutrition, number of childbirths, and heavy labor, were regarded as its etiology. Although some researchers including Dr. Noboru Hagino, the Director of Hagino Hospital, suggested that mine pollution was the culprit, no specific investigations or studies were conducted. Under these circumstances, the Fuchu Town Agricultural Mutual Aid Association requested Dr. Kinichi Yoshioka in August 1960 to investigate the causes of rice growth disorder. He realized that the *itai-itai* disease was occurring endemically and conducted an investigation with the cooperation of Hagino. He published the "Report of Mineral Hazard Downstream of the Jinzu River and its Tributaries—Mineral Hazard on Agriculture & on Human Being (*Itai-itai* Disease)" [4] in June 1961. This was the first report that indicated the possibility of Cd poisoning stating that "*itai-itai* disease is a chronic heavy metal poisoning (mainly Cd), and this report has an international significance due to the fact that it occurred in paddy farming zones that receive the outflow from Cd-producing ores" [4].

This report prompted Toyama Prefecture and the national government to conduct studies, which started between 1962 and 1963. Based on the findings, the Ministry of Health and Welfare published the following opinion in May 1968: "*Itai-itai* disease is a chronic poisoning of Cd that is excreted by the Kamioka Mine and is a pollution disease" [5]. Various Cd environmental pollution measures were continuously undertaken thereafter, such as the enactment of the "Act on Special Measures concerning Relief of Health Damage Related to Pollution" (1969), the enactment of the "Anti-Farm Soil Pollution Law" (1970) that mandates the allowable Cd range in rice crops to be <1 ppm, and the subsequent measures to restore contaminated paddy field soil (1972).

As such, half a century has passed since 1968, when the Ministry of Health and Welfare acknowledged *itai-itai* disease as a pollution disease. This chapter does not provide an overview of the entire history of studies on *itai-itai* disease but rather focused on the following three topics, which were not explored previously. The first topic is the significance of studies conducted by Dr. Kinichi Yoshioka who published the first report on Cd environmental pollution. The second topic is the summary on the opinions of the Ministry of Health and Welfare, which became the starting point for aforementioned governmental measures. The third topic concerns the trends and problems associated with Japan's Cd studies. Chapter 2 of this book illustrates the epidemiology, current circumstances, and challenges of *itai-itai* disease and Cd nephropathy in the Jinzu River basin of Toyama Prefecture.

1.1.2 Discovery of Cd Environmental Pollution: Dr. Kinichi Yoshioka's Performance Re-evaluation

Yoshioka has criticized medical scientists in his book, "Study of *Itai-itai* Disease," [6] as follows:

> First, the true condition of *itai-itai* disease had not been elucidated by Japanese medicine for a long time because sufficient epidemiological studies, such as that illustrated in 'Report of Mineral Hazard Downstream of the Jinzu River and its Tributaries—Mineral Hazard on Agriculture & on Human Being (*Itai-itai* Disease)' [4] that was published in June 1961, had not been conducted. (Page 145)

> The mine pollution study primarily elucidates the route in which heavy metals, including Zn, Pb, and Cd, flow down from mines concentrated in certain areas of the earth's surface, thereby damaging the health of plants and animals in the affiliated regions. This requires a comprehensive knowledge about various fields of sciences, including natural science, technology science, and social science, ranging from geomorphology, geology, mining and metallurgy, to riverology, irrigation, soil science, and chemistry, as well as physiology, ecology, pathology and the hygiene study of animals and plants, as well as political economics. The study planner must also systematically conduct sampling in the target region. (Page 12)

In this way, Yoshioka conducted a systematic sampling of the minerals, river water, agricultural water, and paddy soil to elucidate the relation between Jinzu River's water system and mining pollution. The drainage from the contamination source, including the Kamioka Mine, smelter and waste dump, as well as river water, rice paddy soils (categorized by soil layers, water inlets, and outlets), plants (barnyard grass, horse chestnut, mugwort, astragalus), crops (rice, Chinese cabbage, radish, spinach, persimmon), fish (crucian carp, Japanese dace, carp, loach, masu trout, sand borer), and chicken eggs, was sampled. Hair, blood, and organ samples from patients suffering from *itai-itai* disease were also sampled; the concentrations of Cd, lead, and zinc from each sample were measured. A control sample survey was also simultaneously conducted to clarify the existence of high-concentration contamination of heavy metals confined to the Jinzu River water system and the route of exposure. When considering exposure to human beings due to heavy metal environmental pollution, the actual condition of environmental pollution and the route to human exposure must be systematically clarified. Yoshioka was the first to conduct such ecological and epidemiological studies to determine the cause of *itai-itai* disease. Unfortunately, there have been very few ecological and epidemiological studies [7] such as the one conducted by Yoshioka. Moreover, although Yoshioka reported the results of social and economic investigations [4, 8], most other studies have not conducted such investigations on *itai-itai* disease.

1.1.3 Background Information and Significance of Environmental Administration that Led to the Formulation of the Ministry of Health and Welfare's Opinion: Retrospection from the Viewpoint of Regulatory Science

Based on Yoshioka's report [4], Hagino and Yoshioka announced the possibility of Cd poisoning in their report titled "Study on the Causes of *Itai-itai* Disease" [9] at The 34th General Meeting of the Japanese Orthopedic Association in June 1961. The announcement at the association became the turning point, which inspired Toyama Prefecture and the national government (Ministry of Health and Welfare and the Ministry of Education, Science, Sports and Culture) to organize a research group after the period from 1962 to 1963. The research group conducted additional tests with regard to Yoshioka's study [10, 11]. A notable example is the research group (group leader: Dr. Itsuzo Shigematsu) on "causes of the *itai-itai* disease, particularly the distribution and origin of heavy metals (mainly Cd)" that was funded by a pollution research investigation grant from the Ministry of Health and Welfare in 1967, which measured the concentration of Cd, zinc, and lead from 564 samples, including the waste water from the Kamioka Mine, the mud immediately beneath said mine, as well as river water, rice paddy soil, and rice, including waste slag from the mine's dumping site. Based on these results, the group concluded that the "excretions from the mine and related facilities are the primary cause of these heavy metals" [11].

The Ministry of Health and Welfare published the following opinion on May 8, 1968, to summarize the progress of their investigative study: "*Itai-itai* disease is a chronic poisoning of Cd that is excreted by the Kamioka Mine and is a pollution disease" [5]. Mr. Michio Hashimoto, an administrative official from the Ministry of Health and Welfare, who practically summarized the opinions of the said organization, wrote the following in his book [12]:

> I felt that the report (refer to Literature 10 (original author's note)) published by the joint research group of the Ministry of Health & Welfare and Ministry of Education indicating Cd as a necessary factor in causing *itai-itai* disease should be utilized by the pollution administration in making a decision regarding pollution administration since there remains so much uncertainty. (Page 129)

> I believed that it would be impossible to seek a more proactive decision than the one made by the joint research group of the Ministry of Health & Welfare and Ministry of Education, wherein "it is considered a factor," even after requesting the researchers to scientifically prove the causal relation with Cd using the investigation and study results reported thus far. This lesson was learned from the bitter experience of Kumamoto Minamata disease, in which the causal investigation committee of the Ministry of Health & Welfare, Ministry of

International Trade and Industry, Economic Planning Agency had discontinued their investigation to determine the causal relation simply because the researchers had different opinions. Therefore, I decided to summarize the opinions of the pollution administration based on scientific information, such as studies and investigation reports, concerning *itai-itai* disease, officially publish the opinions of the Ministry of Health & Welfare, and promote preventive measures and relief for pollution-related health hazards. These are all summarized in the "The Ministry of Health & Welfare's Opinion on *Itai-itai* Disease & Reference Materials." (Page 133)

The situation confronted by Hashimoto and the administrative judgment raised important issues concerning the ways in which the regulatory science at the time contributed to the decision-making with regard to environmental policies and administrations [13]. The Rio de Janeiro Declaration adopted at the United Nations Conference on Environment and Development in 1992 states that if there is a risk of serious or irreversible damage, then the lack of sufficient scientific certainty shall not be used as a reason to delay the implementation of cost-effective measures that can prevent environmental deterioration [13]. In the case of *itai-itai* disease, "serious damages" were already being incurred. In addition, it took >50 years from the first onset and >10 years from 1955 for *itai-itai* disease to be recognized as a social problem and for the government to take administrative steps. It was a case wherein the government had seriously delayed taking any action. Hashimoto's writing is a frank expression about the mind-set and practices of administrative officials concerning environmental administration; from the viewpoint of developing regulatory science, it raises many issues that should be considered even today.

1.1.4 Trends and Issues Related to Cd Research in Japan

Hashimoto, who summarized the opinions of the Ministry of Health and Welfare, stated that in terms of scientifically proving the causal relation between *itai-itai* disease and Cd, about <50% of the mechanisms are still unknown, and scientific elucidation must be proactively continued hereafter [12]. He also stated that a medical research group for chronic Cd poisoning and *itai-itai* disease was launched in May 1969. Since then, various studies on *itai-itai* disease have been conducted for the past 17 years, including resident health surveys, environmental pollution surveys, Cd poisoning and *itai-itai* disease mechanisms, and differential diagnosis [12].

The aforementioned research group is still active as of 2018 and has been in operation for more than 48 years. A prominent summary of their studies includes "Cadmium Studies in Japan—A Review" [14], published in 1978, which summarizes the progress made until 1976, and "Interim Summary Report on Studies of Health Effects of Cadmium" [15], published in 1989, which summarizes their progress from 1976 to 1988. The following section considers problems that were described in the conclusions of the two aforementioned publications.

1.1.4.1 Regarding "Cadmium Studies in Japan—A Review" [14]

In "Cadmium Studies in Japan—A Review" [14], the editor, Dr. Kenzaburo Tsuchiya, wrote the following introduction, which describes the objective of the publication:

> Although some studies on *itai-itai* disease have been reported in English as well as in Japanese, scientists abroad have not always been able to keep abreast of these studies and investigations. In particular, they are not familiar with the controversy existing among Japanese scientists over the etiology of *itai-itai* disease. (Page xix)

The aforementioned dispute among researchers on the cause of *itai-itai* disease refers to the controversy regarding whether the cause of *itai-itai* disease is Cd. In the studies concerning the formulation of the Ministry of Health and Welfare's opinion, it has been clarified that the cause of *itai-itai* disease is the heavy metals being discharged from the Kamioka Mine, particularly Cd; nevertheless, there remained some skeptical researchers. The editor, Kenzaburo Tsuchiya, is believed to be one of them. Therefore, Chap. 7 "Etiology of *Itai-itai* Disease" and Chap. 8 "Conclusion" in the said book edited by Tsuchiya must be critically reviewed. This review will not be conducted here, but the original author should point out that Tsuchiya could not deny the involvement of Cd and wrote in Chap. 8 "Conclusion" of the said book as follows:

> When all available data are taken into consideration, it is difficult to rule out the possibility of the involvement of cadmium in the development of *itai-itai* disease. (Page 304)

1.1.4.2 Regarding "Interim Summary Report on Studies of Health Effects of Cadmium" [15] and "Cd Environmental Pollution Regional Resident Health Survey" Report [15]

A summary of the outcome of "Comprehensive Research Group of *Itai-itai* Disease & Chronic Cd Poisoning," commissioned by the Ministry of the Environment from 1976 to 1988, was reported in 1989, but the Research Committee (chairman: Dr. Kenzaburo Tsuchiya) published a summary report claiming that "there are still quite a few elements in the onset mechanism of proximal renal tubular dysfunction, and bone lesions seen in *itai-itai* disease patients that have yet to be elucidated" [15]. The original authors conducted a detailed examination of the renal tubular function and bone/calcium/phosphorus metabolism in the Cd nephropathy cases (21 males and 13 females) in the Cd-contaminated area of Jinzu River basin in Toyama Prefecture [16]. Thus, a decrease in tubular phosphorus reabsorption function, decrease in serum inorganic phosphorus concentration, and increase in serum alkaline phosphatase activity were observed in both males and females. It was revealed that hypophosphatemia due to decreased proximal tubular reabsorption of phosphorus was the major pathological condition that was leading to abnormal bone metabolism (Fig. 1.1) [16, 17]. The latest findings on hypophosphatemia related to Cd nephropathy are detailed in Chap. 2.

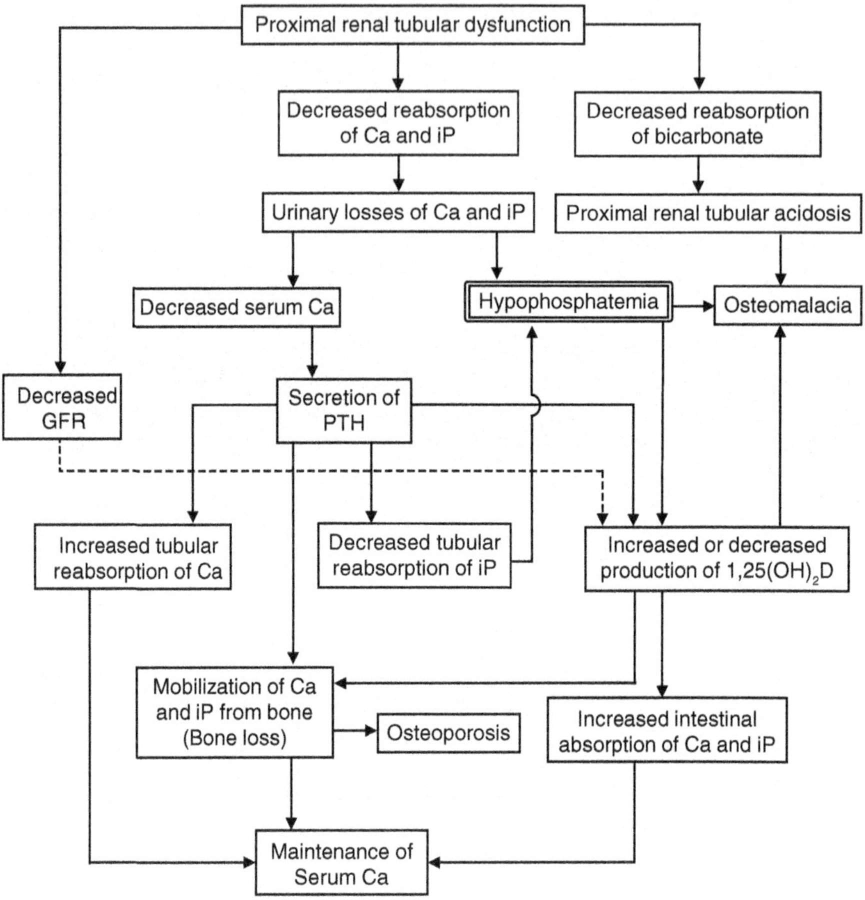

Fig. 1.1 Proposal schema for the pathogenesis of *itai-itai* disease. Solid lines indicate a stimulatory action and the dotted line an inhibitory action. (Aoshima 1997, with permission from Yakugaku Zasshi)

The "Interim Summary Report on Studies of Health Effects of Cadmium" [15] summarized the results of both the "Cd Environmental Pollution Regional Resident Health Survey" conducted in seven prefectures (Akita, Fukushima, Gunma, Ishikawa, Hyogo, Nagasaki, and Oita prefectures) between 1976 and 1987 and the investigation conducted at the Jinzu River basin in Toyama Prefecture between 1979 and 1984. The number of subjects in the Cd-contaminated areas in the 8 prefectures was 8266, 2168, 765, 628, 534, 479, 372, and 358 in Toyama, Gunma, Fukushima, Ishikawa, Hyogo, Akita, Oita, and Nagasaki prefectures, respectively. The results of the age-specific urinary β_2-microglobulin positive rate in each contaminated area of Toyama, Ishikawa, Hyogo, and Nagasaki prefectures as well as the nonpolluted area (control group) of this report are shown in Fig. 1.2 [17]. In all four areas, urinary β_2-microglobulin positive rate showing tubular dys-

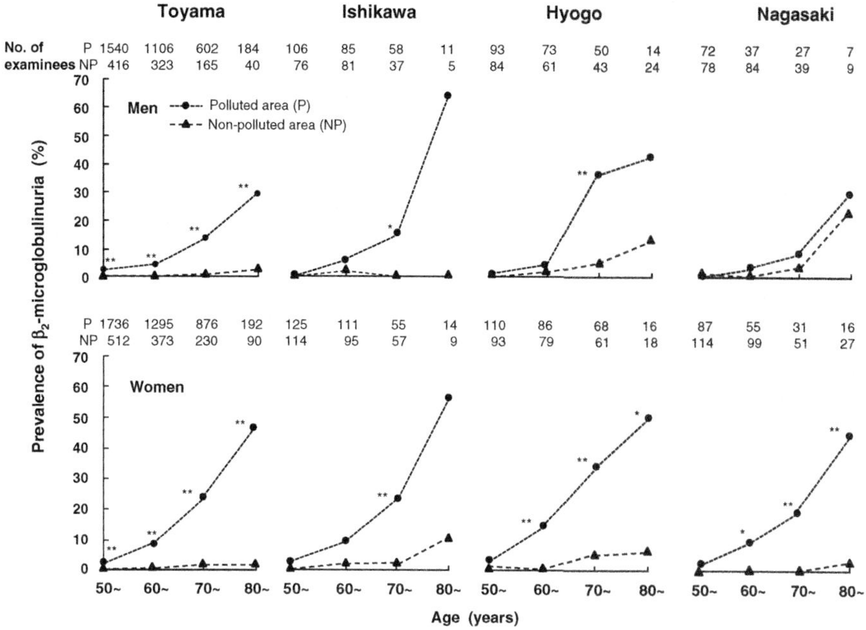

Fig. 1.2 Prevalence of β_2-microglobulinuria in men and women living in cadmium-polluted (P) and non-polluted (NP) areas in four prefectures in Japan. $*p < 0.05$, $**p < 0.01$. (Aoshima 1997, with permission from Yakugaku Zasshi)

function was higher in Cd-contaminated areas than in non-contaminated areas, particularly in females, and the difference was remarkable in older subjects. Furthermore, in the same study, the "presence of proximal renal tubular dysfunction," which is characterized by a tubular phosphorus reabsorption rate of <80% and blood bicarbonate ion concentration of <23 mEq/L, was only observed in the residents of the polluted regions. This condition was applicable to a total of 202 subjects: 162 (Toyama), 16 (Ishikawa), 11 (Hyogo), 12 (Nagasaki), and 1 (Fukushima). This study was conducted to clarify the health effects caused by Cd environmental pollution. As shown in Fig. 1.2, in Toyama, Ishikawa, Hyogo, and Nagasaki Prefectures, there was a clear relation between Cd environmental contamination and the appearance of proximal renal tubular dysfunction. The report claimed that "no correlation is observed in the degree of environmental contamination in the seven prefectures and health effects;" however, there is a problem with using only Cd concentration in rice crops as an indicator of the degree of environmental pollution because it requires detailed consideration and evaluation of geomorphology/geology of the mine, which is the source of contamination, mining method, presence/absence of beneficiation/smelting, operation period, and contamination route to human exposure. Furthermore, the results of this study are introduced in WHO's "Environmental Health Criteria 134: Cadmium" [18]. This book describes the detailed results of surveys on health effects that have been conducted thus far in each Cd-contaminated area: Jinzu River basin in Toyama Prefecture (Chap. 2), Kakehashi River basin in Ishikawa Prefecture (Chap. 4), and Akita Prefecture (Chap. 6) are detailed therein.

1.2 Current Situation of Japanese People's Exposure to Cd

Due to the discovery of *itai-itai* disease in Toyama Prefecture and the involvement of Cd contamination, the interest in Cd exposure in the general population of Japan has increased nationwide. Various nationwide investigations by the national government have been conducted since then.

Regarding Cd intake by the Japanese, the findings of the National Institute of Health Sciences have been published from 1980 to 2015 (Fig. 1.3) [19]. The researchers estimated the Cd intake level based on the food intake level that was elucidated by conducting the National Health and Nutrition Survey, by gathering and cooking Japanese food, and by analyzing the Cd concentration within the foods. According to the study, Cd intake per person per day in 1981 was 31.1 μg, but it continued to decrease despite many fluctuations. Finally, in 2015, it was reduced to 17.8 μg, which decreased >40% in ~30 years.

Thus, the amount of Cd intake from rice was the most significant, among other items such as agricultural products and marine products (Fig. 1.4). Although the Cd intake level from rice was 16.2 μg per person/day in 1981, which accounted for 52.1% of the total Cd intake, it decreased to 7.84 μg (37.2%) by 2007 and 5.67 μg

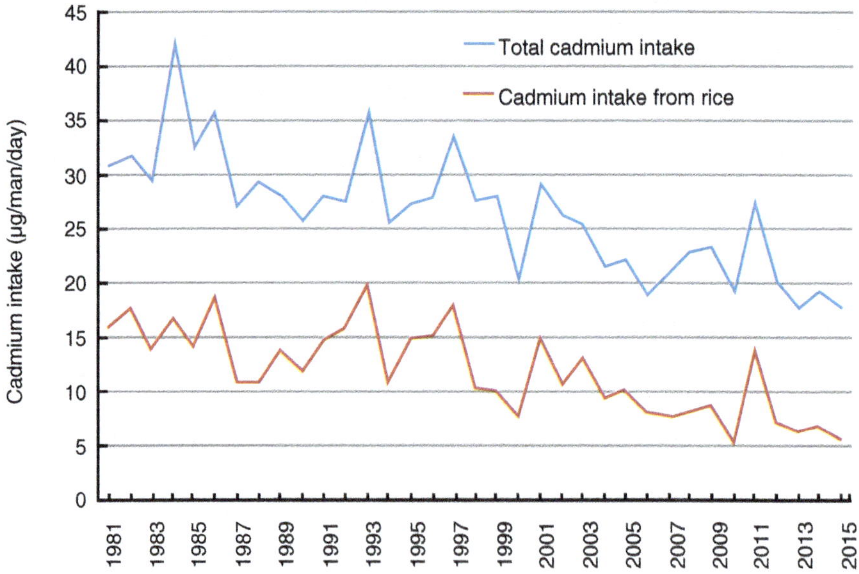

Fig. 1.3 Annual change of cadmium intake in Japanese. Ministry of Agriculture, Forestry and Fisheries of Japan

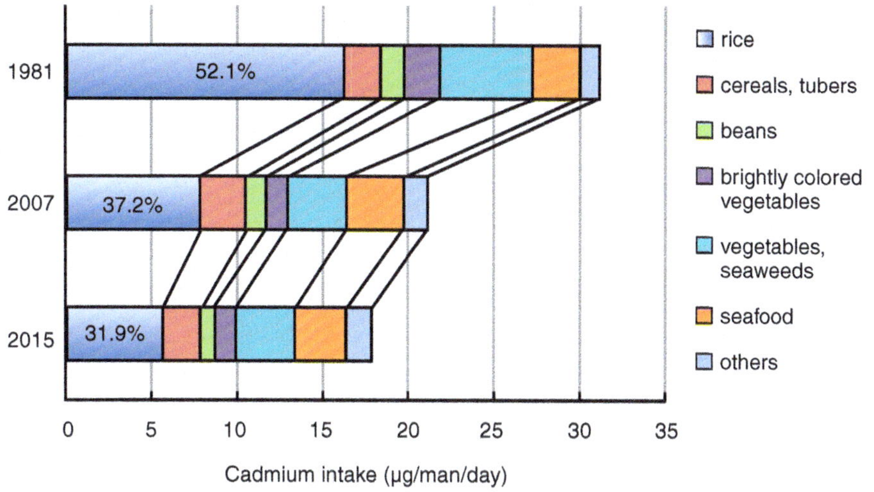

Fig. 1.4 Breakdown of cadmium intake in Japanese. Ministry of Agriculture, Forestry and Fisheries of Japan

(31.9%) by 2015, which shows a significant decrease to approximately one-third the level within ~30 years. In addition, the time-dependent variation pattern of Cd uptake derived from rice is almost parallel to that of the whole Cd intake (Fig. 1.3). Therefore, the Cd intake from rice has seen a significant decrease that eventually decreased the Cd intake of the Japanese.

The rice-derived Cd intake level of the Japanese has decreased due to two possible reasons. According to the food balance sheet published by the Ministry of Agriculture, Forestry and Fisheries [20], the consumption of rice per person had decreased almost consistently from 118.3 kg in 1962, which was the peak time, and particularly between 1981 and 2015, when it decreased by 2/3 from 77.8 to 54.6 kg in ~30 years (Fig. 1.5). Second, countermeasures have been taken to reduce the concentration of Cd in rice, such as via flood control. The Ministry of Agriculture, Forestry and Fisheries conducted the "Survey of Cadmium in Domestic Agricultural Products" to determine the effects of technology dissemination in reducing the Cd concentration in domestic agricultural products [21]. The survey entailed sampling rice in municipalities across the country and investigating the Cd concentrations. The survey was conducted in 1997 and 1998, and after flood control had been implemented, it was conducted again in 2009 and 2010. The results showed that the Cd concentrations in rice had been reduced, and rice with a Cd concentration exceeding the reference value of 0.4 mg/kg was no longer detected (Table 1.1). In addition, the Ministry of Agriculture and Forestry conducted a "monitoring survey of rice in production areas" between 1999 and 2010 to confirm the effect of the reduction measures during the production stage in areas with relatively high Cd concentrations [22]. According to the survey, since 2003, rice product with a Cd concentration exceeding the reference value was almost nonexistent in areas wherein absorption-suppressing measures focused on flood management had been implemented. This coincides with the time when measures

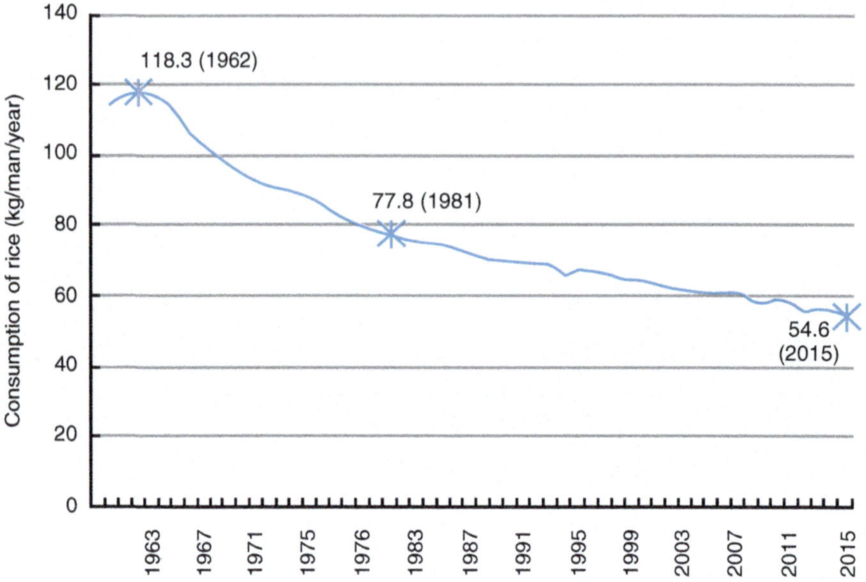

Fig. 1.5 Annual change of consumption of rice in Japanese. Ministry of Agriculture, Forestry and Fisheries of Japan

Table 1.1 Comparison of distribution of rice cadmium concentration between before and after prosecuting measures to decrease rice cadmium concentration in Japan

Year	1997, 1998	2009, 2010
Number	37,250	2000
≤0.1 mg/kg	30,346 (81.5%)	1790 (89.5%)
0.1 mg/kg<, ≤0.2 mg/kg	5457 (14.6%)	175 (8.8%)
0.2 mg/kg<, ≤0.3 mg/kg	1101 (3.0%)	32 (1.6%)
0.3 mg/kg<, ≤0.4 mg/kg	250 (0.7%)	3 (0.2%)
0.4 mg/kg<	96 (0.3%)	0 (0.0%)
Mean (mg/kg)	0.06	0.05
25 percentile (mg/kg)	0.02	0.02
Median (mg/kg)	0.04	0.02
75 percentile (mg/kg)	0.08	0.06
90 percentile (mg/kg)	0.14	0.10
95 percentile (mg/kg)	0.18	0.14

Ministry of Agriculture, Forestry and Fisheries

to suppress Cd absorption had been promoted and implemented (Table 1.2). Therefore, the Cd intake of the Japanese people decreased due to the additive effect of the reduction of rice intake and the decrease in concentration of Cd in rice.

However, the amount of Cd intake from foods other than rice has not shown any significant change (Fig. 1.3). Among them, Cd intake from agricultural products has been trending slightly downward (Fig. 1.4). This trend has also been confirmed

Table 1.2 Annual change of distribution of rice cadmium concentration in the areas where rice with relatively high cadmium concentration was produced in Japan (%)

Year	1999	2000	2001	2002	2003	2004	2005	2006	2007	2008	2009	2010
Number	1936	508	515	945	2740	2281	1931	1190	1065	926	30	197
1.0 mg/kg<	0.2	0.0	0.4	0.1	0.0	0.0	0.1	0.0	0.0	0.0	0.0	0.0
0.4 mg/kg<, ≤1.0 mg/kg	3.3	9.3	6.4	3.2	1.6	0.8	1.1	0.3	0.2	0.3	0.0	0.5
≤0.4 mg/kg	96.5	90.7	93.2	96.7	98.4	99.2	98.9	99.7	99.8	99.7	100.0	99.5
Total	100.0	100.0	100.0	100.0	100.0	100.0	100.0	100.0	100.0	100.0	100.0	100.0

Ministry of Agriculture, Forestry and Fisheries

Table 1.3 International comparison of food supply in 2013 (kcal/capita/day)

	Rice	Fresh-water fish	Demersal fish	Pelagic fish	Marine fish, others	Crusta-ceans	Cephalo-pods	Molluscs, others	Aquatic plants
Japan	646	11	39	51	15	4	8	4	1
USA	72	8	9	9	0	6	1	1	0
Canada	135	10	8	12	0	5	0	2	0
UK	68	6	15	13	0	4	0	1	0
France	52	8	37	18	1	5	1	4	0
Germany	35	7	20	7	1	1	0	0	0
Italy	59	5	13	17	2	2	6	2	0
Spain	85	6	23	35	2	7	7	6	0
Russian Federation	45	12	22	17	0	1	2	0	0

Food and Agriculture Organization

in the aforementioned "Survey of Cadmium in Domestic Agricultural Products" [21]. In contrast, although the amount of Cd intake from fish has hardly changed, the proportion of Cd intake derived from fish and shellfish was 8.7% of the total Cd intake in 1985, which doubled to 16.9% by 2015. Furthermore, since data on sea weed is included in the same category as vegetables in the survey conducted by the National Institute of Health Sciences, the ratio is expected to be even higher when calculating it based on the category of "seafood" that includes the data on seaweed. Therefore, seafood has attained an important status alongside rice and other agricultural products as a source of Cd in the Japanese diet in recent years.

In this way, although the Cd intake of the Japanese people is on a downward trend, the amount of intake of rice and marine products is large compared with that of Western countries (Table 1.3) [23]. Therefore, the Cd intake in Japan is estimated to be at a relatively high level internationally. According to various epidemiological studies conducted in non-Cd polluted areas both in Japan and abroad, in many cases the Cd concentrations in the blood and urine of Japanese people tend to be higher than those in Western countries (Table 1.4). This reflects the high intake of Cd derived from rice and marine products that constitute a large proportion of the Japanese diet.

Table 1.4 Blood and urinary cadmium levels in people nonpolluted by cadmium from various countries

Subjects	Sex	Age	Number	Blood cadmium	Urinary cadmium	Year	References
General population in Japan	F	20–86	11,090	–	1.32 (μg/g cr.) (GM)	2000–2002	[24]
General population in Japan	F	20–81	1420	1.23 (μg/L) (GM)	–	2002, 2008	[25]
General population in rural areas of Japan	M	50–99	1105	–	1.8 (μg/g cr.) (GM)	1993, 1994	[26]
	F	50–99	1611	–	2.4 (μg/g cr.) (GM)		
Japanese farmers in a nonpolluted area	F	40–69	187	2.00 (μg/L) (GM)	2.63 (μg/g cr.) (GM)	2001	[27]
Japanese farmers in a nonpolluted area	F	34–74	123	2.21 (μg/L) (GM)	3.36 (μg/g cr.) (GM)	2003	[28]
Population in an industrial area of Belgium (CadmiBel)	M	≥20	106	6.5 (nmol/L) [0.73 (μg/L)] (GM)	0.50 (μg/g cr.) (GM)	1985–1989	[29, 30]
	F	≥20	165	6.2 (nmol/L) [0.70 (μg/L)] (GM)	0.55 (μg/g cr.) (GM)		
Population in an rural area of Belgium (CadmiBel)	M	≥20	151	10.0 (nmol/L) [1.12 (μg/L)] (GM)	0.59 (μg/g cr.) (GM)		
	F	≥20	197	11.1 (nmol/L) [1.25 (μg/L)] (GM)	0.75 (μg/g cr.) (GM)		
General population in Sweden (WHILA)	F	53–64	715 (blood) 790 (urine)	0.38 (μg/L) (MD)	0.76 (μg/g cr.)	1999–2000	[31]
Population in a farm area of Sweden	M&F	14–73	105	1.8 (nmol/L) [0.20 (μg/L)] (MD)	0.23 (nmol/mmol cr.) [0.23 (μg/g cr.)] (MD)	–	[32]

Table 1.4 (continued)

Subjects	Sex	Age	Number	Blood cadmium	Urinary cadmium	Year	References
General population in Spain	M&F	15–85	1397	–	0.39 (μg/g cr.) (MD)	1997–2003	[33]
General population in the USA (NHANES)	M	≥20	7587	0.39 (μg/L) (MD)	–	1999–2006	[34]
	F	≥20	7191	0.40 (μg/L) (MD)	–	1999–2006	[35]

F female, *M* male, *GM* geometric mean, *MD* median
To convert blood cadmium from μg/L to nmol/L, multiply by 8.897
To convert urine cadmium from μg/g creatinine to nmol/mmol creatinine, multiply by 1.006

References

1. Saito H, Aoshima K. Heavy metals & human body II - the impact of heavy metals on human body. In: Chino M, Saito H, editors. Heavy metals & biology. Tokyo: Hakuyusha Publishers, Ltd; 1988. p. 167–211.
2. Friberg L, Elinder CG, Kjellstrom T, Nordberg GF. Cadmium and health: a toxicological and epidemiological approach, vol. Vol 1, 2. Boca Raton: CRC Press; 1986.
3. Aoshima K, Iwata K, Kasuya M. Environmental exposure to cadmium and effects on human health. Prat 1. Renal tubular function in inhabitants of the cadmium-polluted Jinzu River basin in Toyama prefecture. Jpn J Hyg. 1988;43:853–63. (in Japanese with English abstract)
4. Yoshioka K. Report of mineral hazard downstream of the Jinzu River and its tributaries—mineral hazard on agriculture and on human being (Itai-itai disease). 1961. (in Japanese). In: Yoshioka K, editor. Study of Itai-itai Disease. Yonago: Tatara Shoboh Publisher, Ltd; 1970. p. 5–95.
5. Ministry of Health and Welfare. Opinion of the Welfare Ministry with regard to itai-itai disease and its causes in Toyama Prefecture. Collection of Special Documents about the Pollution of Government Office. 1968;3:12–25. (in Japanese)
6. Yoshioka K. Study of Itai-itai Disease. Yonago: Tatara Shoboh Publisher, Ltd; 1970. p. 1–334. (in Japanese with English summary)
7. Aoshima K. Itai-itai disease: lessons from the investigations of environmental epidemiology conducted in the 1970's, with special reference to the studies of the Toyama Institute of Health. Jpn J Hyg. 2017;72:149–58. (in Japanese with English abstract)
8. Yoshioka K. Epidemiological study on relationship between itai-itai disease and mining nuisance. Yamaguchi Igaku. 1964;13:146–71. (in Japanese)
9. Hagino N, Yoshioka K. A study on the cause of itai-itai disease. J Jpn Orthop Assoc. 1961;35:812–5. (in Japanese)
10. Toyama Prefecture. Report of the Regional Special Disease Control Committee. 1967. (in Japanese)
11. Japan Public Health Association. Study on the Etiology of Itai-itai Disease with Special Reference to the Distribution and Origins of Heavy Metals, Especially Cadmium; 1968. p. 1–58. (in Japanese)
12. Hashimoto M. My personal history with environmental administration. Tokyo: Asahi Newspapers; 1988. p. 1–376. (in Japanese)

13. Environmental Risk Subcommittee, a joint venture by Health & Life Science Committee and Environmental Studies Committee, Science Council of Japan. The State of Regulatory Science on Decision Making in Environmental Administration (Report); 2017. p. 1–24. (in Japanese)
14. Tsuchiya K. Cadmium studies in Japan: a review. New York: Kodansha; 1978. p. 1–376.
15. Japan Public Health Association Cadmium Research Committee. Interim summary report on studies of health effects of cadmium. Kankyo Hoken Report. 1989;56:1–345. (in Japanese with English abstract)
16. Aoshima K, Katoh T, Teranishi H, Horiguchi H, Kasuya M. Abnormalities of calcium, phosphorus and vitamin D metabolism with proximal renal tubular dysfunction in subjects environmentally exposed to cadmium. Jpn J Hyg. 1993;47:1009–20. (in Japanese with English abstract)
17. Aoshima K. Recent advances in studies of itai-itai disease. Jpn J Toxicol Environ Health. 1997;43:317–30. (in Japanese with English abstract)
18. World Health Organization (WHO). Environmental health criteria 134. cadmium. Geneva: WHO; 1992.
19. Ministry of Agriculture, Forestry and Fisheries of Japan. The status of cadmium concentrations in agricultural productions and intake of cadmium in Japan. http://www.maff.go.jp/j/syouan/nouan/kome/k_cd/jitai_sesyu/01_inv.html.
20. Ministry of Agriculture, Forestry and Fisheries of Japan. Food balance sheet. http://www.maff.go.jp/j/zyukyu/fbs/.
21. Ministry of Agriculture, Forestry and Fisheries of Japan. Surveys of actual status of cadmium concentrations in agricultural products. http://www.maff.go.jp/j/press/syouan/nouan/160223.html.
22. Ministry of Agriculture, Forestry and Fisheries of Japan. Monitoring of cadmium concentration of rice in rice-producing areas of Japan. http://www.maff.go.jp/j/syouan/nouan/kome/k_cd/documents/02_mon.html.
23. Food and Agriculture Organization. FAOSTAT. Food balance sheets. http://www.fao.org/faostat/en/#data/FBS
24. Moriguchi J, Ezaki T, Tsukahara T, Fukui Y, Ukai H, Okamoto S, Shimbo S, Sakurai H, Ikeda M. Effects of aging on cadmium and tubular dysfunction markers in urine from adult women in non-polluted areas. Int Arch Occup Environ Health. 2005;78:446–51.
25. Ikeda M, Ohashi F, Fukui Y, Sakuragi S, Moriguchi J. Cadmium, chromium, lead, manganese and nickel concentrations in blood of women in non-polluted areas in Japan, as determined by inductively coupled plasma-sector field-mass spectrometry. Int Arch Occup Environ Health. 2011;84:139–50.
26. Suwazono Y, Kobayashi E, Okubo Y, Nogawa K, Kido T, Nakagawa H. Renal effects of cadmium exposure in cadmium nonpolluted areas in Japan. Environ Res. 2000;84:44–55.
27. Horiguchi H, Oguma E, Sasaki S, Miyamoto K, Ikeda Y, Machida M, Kayama F. Dietary exposure to cadmium at close to the current provisional tolerable weekly intake does not affect renal function among female Japanese farmers. Environ Res. 2004;95:20–31.
28. Horiguchi H, Aoshima K, Oguma E, Sasaki S, Miyamoto K, Hosoi Y, Katoh T, Kayama F. Latest status of cadmium accumulation and its effects on kidneys, bone, and erythropoiesis in inhabitants of the formerly cadmium-polluted Jinzu River Basin in Toyama, Japan, after restoration of rice paddies. Int Arch Occup Environ Health. 2010;83:953–70.
29. Sartor FA, Rondia DJ, Claeys FD, Staessen JA, Lauwerys RR, Bernard AM, Buchet JP, Roels HA, Bruaux PJ, Ducoffre GM, Lijnen PJ, Thijs LB, Amery AK. Impact of environmental cadmium pollution on cadmium exposure and body burden. Arch Environ Health. 1992;47:347–53.
30. Sartor F, Rondia D, Claeys F, Buchet JP, Ducoffre G, Lauwerys R, Staessen J, Amery A. Factors influencing the cadmium body burden in a population study. IARC Sci Publ. 1992;118:101–6.
31. Suwazono Y, Sand S, Vahter M, Filipsson AF, Skerfving S, Lidfeldt J, Akesson A. Benchmark dose for cadmium-induced renal effects in humans. Environ Health Perspect. 2006;114:1072–6.

32. Akesson A, Bjellerup P, Lundh T, Lidfeldt J, Nerbrand C, Samsioe G, Skerfving S, Vahter M. Cadmium-induced effects on bone in a population-based study of women. Environ Health Perspect. 2006;114:830–4.
33. Olsson IM, Bensryd I, Lundh T, Ottosson H, Skerfving S, Oskarsson A. Cadmium in blood and urine--impact of sex, age, dietary intake, iron status, and former smoking--association of renal effects. Environ Health Perspect. 2002;110:1185–90.
34. Navas-Acien A, Tellez-Plaza M, Guallar E, Muntner P, Silbergeld E, Jaar B, Weaver V. Blood cadmium and lead and chronic kidney disease in US adults: a joint analysis. Am J Epidemiol. 2009;170:1156–64.
35. Grau-Perez M, Pichler G, Galan-Chilet I, Briongos-Figuero LS, Rentero-Garrido P, Lopez-Izquierdo R, Navas-Acien A, Weaver V, García-Barrera T, Gomez-Ariza JL, Martín-Escudero JC, Chaves FJ, Redon J, Tellez-Plaza M. Urine cadmium levels and albuminuria in a general population from Spain: A gene-environment interaction analysis. Environ Int. 2017;106:27–36.

Part II
Health Effects of Cadmium in Human Populations and its Relation to Cadmium Exposure Levels

Chapter 2
Recent Clinical and Epidemiological Studies of *Itai-Itai* Disease (Cadmium-Induced Renal Tubular Osteomalacia) and Cadmium Nephropathy in the Jinzu River Basin in Toyama Prefecture, Japan

Keiko Aoshima

Abstract The majority of studies on the biological effects of exposure to the heavy metal cadmium (Cd) have focused on the epidemiology, mechanisms of toxicity, pathophysiology, disease progression, and prognosis of multiple proximal tubular dysfunction (Cd nephropathy). Indeed, this was the earliest symptom to occur in the residents of the downstream Jinzu River basin in Toyama Prefecture, which was contaminated by Cd. Despite efforts to reduce Cd exposure through rice grown in private rice paddies and the replacement of contaminated rice paddy soil, Cd nephropathy has continued to persist in this area. In Cd nephropathy, even when the disease is relatively very mild with fractional excretion (FE) of β_2-microglobulin (β_2-MG) levels of <1 %, FEs of uric acid, calcium, sodium, and chloride increase. The results of a 27-year observational study have indicated that cases presenting with urinary β_2-MG levels of ≥ 10 mg/g Cr have decreased tubular reabsorption of phosphate, decreased glomerular filtration rates (GFRs), and anemia with Cd nephropathy observed at GFRs of <30 mL/min. In addition, as recent as 2015, one case with urinary β_2-MG levels of ≥ 10 mg/g Cr developed osteomalacia and was officially certified as having *itai-itai* disease. Fibroblast growth factor 23 is not involved in the pathogenesis of hypophosphatemia in *itai-itai* disease and Cd nephropathy.

The original version of this chapter was revised. A correction to this chapter can be found at https://doi.org/10.1007/978-981-13-3630-0_15

K. Aoshima (✉)
Hagino Hospital, Toyama, Japan
e-mail: keiaoshima@po4.nsk.ne.jp

S. Himeno, K. Aoshima (eds.), *Cadmium Toxicity*, Current Topics in Environmental Health and Preventive Medicine,
https://doi.org/10.1007/978-981-13-3630-0_2

Keywords Chronic cadmium poisoning · Fibroblast growth factor 23 Hypophosphatemia · *Itai-itai* disease · β_2-microglobulin · Multiple proximal tubular dysfunction · Osteomalacia

2.1 Introduction

Although trace amounts of the heavy metal cadmium (Cd) are present in the human body, it is known to be toxic and is yet to be proven necessary in humans [1, 2]. Cd does not exist in the pure state in nature and is predominantly found in zinc ores. Cd-polluted areas surrounding factories, as well as paddy soil used for agriculture via river water as a result of zinc ore and metal refinery operations, have been reported throughout Japan [1, 2]. Exposure to Cd by drinking contaminated river water and consuming crops raised in contaminated growing fields has caused outbreaks of chronic Cd poisoning in Japan. *Itai-itai* disease (meaning "*it hurts, it hurts*" disease), breakouts of which have repeatedly occurred in the Jinzu River basin in Toyama Prefecture, is the most severe type of chronic Cd poisoning. *Itai-itai* disease is pathogenetically referred to as renal tubular osteomalacia, which is induced by proximal tubular dysfunction (Cd nephropathy) as a result of exposure to environmental Cd [1–3]. New cases of *itai-itai* disease continue to be identified, indicating that it is not a health hazard of the past, but remains a current social and medical problem in Japan [4–6]. There have been many cases of Cd nephropathy caused by chronic Cd poisoning in Toyama Prefecture, and these cases range from mild to as severe as *itai-itai* disease. Herein, the pathophysiology, disease progression, and prognosis of Cd nephropathy are presented based on the evaluation of proximal tubular function testing among residents of Cd-contaminated areas in the Jinzu River basin indicating proximal tubular dysfunction ranging from mild to severe forms. In addition, we investigated whether the fibroblast growth factor 23 (FGF-23), a recently discovered hormone that is produced by osteocytes and that regulates phosphate and vitamin D metabolism [7], is involved in the pathogenesis of hypophosphatemia in *itai-itai* disease and Cd nephropathy. The present chapter may contribute to an increased understanding of the pathophysiology, clinical course, and prognosis of Cd nephropathy and *itai-itai* disease.

2.2 History and the Recent Social Situation of *Itai-Itai* Disease and Cd Nephropathy

The geographical features of the Jinzu River basin, the endemic area of *itai-itai* disease, its industrial hazard history, and the history of the study on *itai-itai* disease have been discussed in a previous review article [4]. Briefly, the first Cd-contaminated area discovered in the world was the Jinzu River basin in Toyama Prefecture, Japan (Fig. 2.1). Toyama is located along the northwest coast of Japan's main island. The Jinzu River flows northward for 120 km to Toyama Bay of the Japan Sea. Two major branches, the Miya and Takahara rivers, flow together into the Jinzu River at the

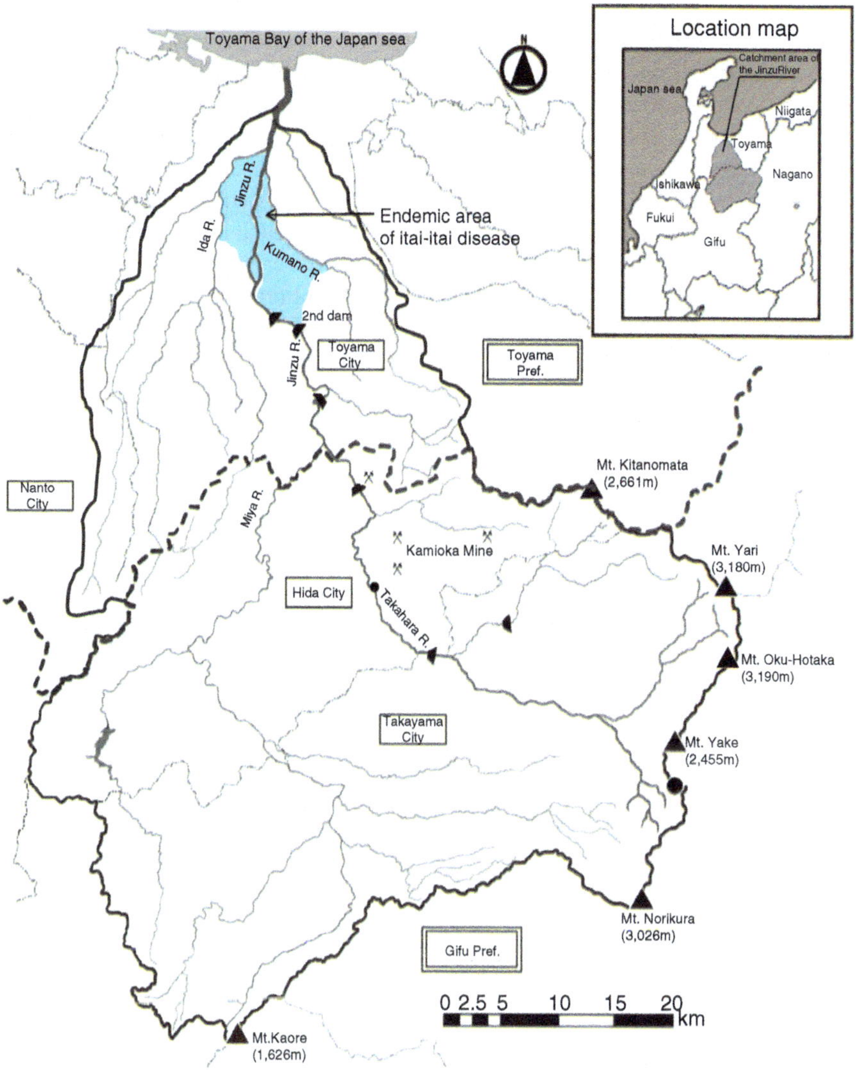

Fig. 2.1 The catchment area of the Jinzu River and the endemic area of *itai-itai* disease in the Jinzu River basin, Toyama Prefecture, Japan. The solid line indicates the catchment area of the Jinzu River, which is also indicated by a shadow area on the location map in the right panel. The dashed line indicates the border of the prefecture. *R.* river, *Mt.* mountain

boundary of Gifu and Toyama prefectures. The Takahara River originates in the 3000-meter mountain plateaus of Mount Yari (elevation 3180 m) and Mount Oku-Hotaka (elevation 3190 m), which are part of the northern Japan Alps. This results in a steep, rapidly flowing river. As winter snowfall is abundant and the annual rainfall is more than 3000 mm, the Takahara River supplies copious water volume to the Jinzu River as it runs through a steep valley. At the Jinzu second dam, the river reaches gentler slopes where a fan-shaped plain has formed. This Jinzu River irrigation area has been a rice cultivation zone since the Edo era in 1633 due to efforts such as rice paddy development and canal excavation [6]. Along the Takahara River, the large Kamioka lead and zinc mine and smeltery (the Kamioka Mining Station of Mitsui Mining & Smelting Co., Ltd.) (Fig. 2.1) has been in operation since 1886. From around 1900, the river water became cloudy and whitish, *ayu* fish and carp began to die, and rice ceased to grow in the associated rice paddies. The inhabitants attributed these incidents to mining-associated pollution originating from the Kamioka Mining Station upstream and protested at local public offices and the mining company, as the water from the Jinzu River was used for agriculture, household use, and drinking [6]. Consequently, the mining company built a settling pond, but the pond was such a meager and makeshift effort that most of the slag continued to flow into the river. In the 1920s, the mine stepped up the production and slag continued to increase. At the same time, the number of inhabitants who suffered pain in various parts of their body, and finally were unable to move or were bedridden, began to increase. Thus, *itai-itai* disease first appeared in the early 1900s, with a more frequent incidence observed from around 1935 to 1960.

Restoration of Cd-contaminated paddy soil was conducted over 33 years from 1979 until completion in 2011 [4, 5]. In response, an "agreement" was executed in December 2013 that confirmed the complete resolution between Mitsui Mining & Smelting Co., Ltd., the causative company, and the victims. The contents of the agreement were:

1. Mitsui Mining & Smelting apologized to the victims.
2. Confirmation that problems related to soil contamination and agricultural damage were resolved by completion of contaminated farmland restoration, with continued response to ongoing investigations from victim group liaison councils that compensate patients and prevent pollution.
3. Establishment of a healthcare support system for residents of the Jinzu River basin, with a one-time lump sum payment (600,000 yen) to patients with renal disorders caused by Cd [in fact, renal disorders were defined as urinary β_2-microglobulin (β_2-MG) levels exceeding 5 mg/g creatinine (Cr), original author's note].
4. Victims will not make future claims, and Mitsui Mining & Smelting will pay the settlement fee in accordance with the complete resolution.

Both parties confirm that the issues have been fully resolved [6, 8]. Although kidney damage caused by Cd has not been recognized by the Ministry of the Environment, Japan as a pollution-related illness, the admission by Mitsui Mining & Smelting on the damage and their compensation to people with renal impairment represented a great achievement for both patients and victims.

2.3 Epidemiology of *Itai-Itai* Disease and Cd Nephropathy

The total number of patients with *itai-itai* disease who have been officially certified by the Act on Compensation of Pollution-Related Health Damage that has been in place in the Toyama Prefectural Government since 1967 is 200 (females,

195; males, 5) as of July 2017. The total number of individuals considered in need of regular observation at risk for the development of osteomalacia, after excluding the number of officially certified patients, is 118 (females, 100; males, 18). Survivors include five officially certified patients (females, three; males, two) and four individuals (females, three; male, one) in need of regular observation. Since 2000, 28 individuals (females, 25; males, 3) have applied for official certification, with 17 of these (females, 15; males, 2) getting subsequently certified.

The following is a description of the recent epidemiological and clinical features of the 19 female individuals born between 1910 and 1933 who applied for certification between 2000 and 2008 [5]. Among the 19 patients, 5 individuals (26%) had been exposed to Cd after 1930–1936 when they first lived in the Cd-contaminated area, indicating that these 5 individuals developed *itai-itai* disease as a result of exposure to Cd at least after 1936. For the 10 of the 19 individuals, the time period between the first appearance of the disease and their first medical examination at Hagino Hospital was at least 20 years, indicating extremely chronic disease progression. All individuals presented with abasia or dysbasia because of extreme pain in their lower extremities at their first medical examination in Hagino Hospital. Five of those individuals had extremely severe symptoms, making them unable to voluntarily move even on the bed, and were determined to be in the severe stage of *itai-itai* disease as per the severity classification system derived by Nakagawa [9] in 1960. All 19 individuals had multiple proximal tubular dysfunction demonstrating β_2-MGuria, glucosuria, uricosuria, and phosphaturia as well as decreased glomerular filtration rate (GFR) that ranged from moderate to severe (4–27 mL/min). Thus, all individuals presented with chronic renal failure due to proximal tubular dysfunction. Eight patients had typical and severe osteomalacia accompanied by radiological signs of pseudofractures (Looser's zones) in the femur, pelvis, and ribs (Fig. 2.2).

A recent epidemiological survey of kidney disease among residents of the Cd-contaminated Jinzu River basin reported many cases of Cd nephropathy. Specifically, according to the Survey on the Health Effects of Residents of the Cd-Contaminated Areas of the Jinzu River basin publicized by the Ministry of the Environment and the Toyama Prefectural Government in August 2009, regular testing of urinary β_2-MG using quantitative immunological method began as part of health examinations [10]. Examination of β_2-MG levels at the first examination in a total of 5861 individuals (females, 3178; males, 2683) over 50 years of age who underwent primary health examinations between 1997 and 2007 indicated that 16% of these individuals (931 individuals; 17% females, $n = 528$ and 15% males, $n = 403$) exceeded 1 mg/g Cr, which is the designated screening level for proximal tubular dysfunction. Five percent, or nearly 300 individuals (297 individuals: 6% females, $n = 197$, and 4% males, $n = 100$), tested at $\geq$10 mg/g Cr, which is the level required for a clinical diagnosis of proximal tubular dysfunction. *Itai-itai* disease was identified in a large number of females, as described above; however, Cd nephropathy was identified in multiple cases of both males and females, indicating that there is no gender-based difference.

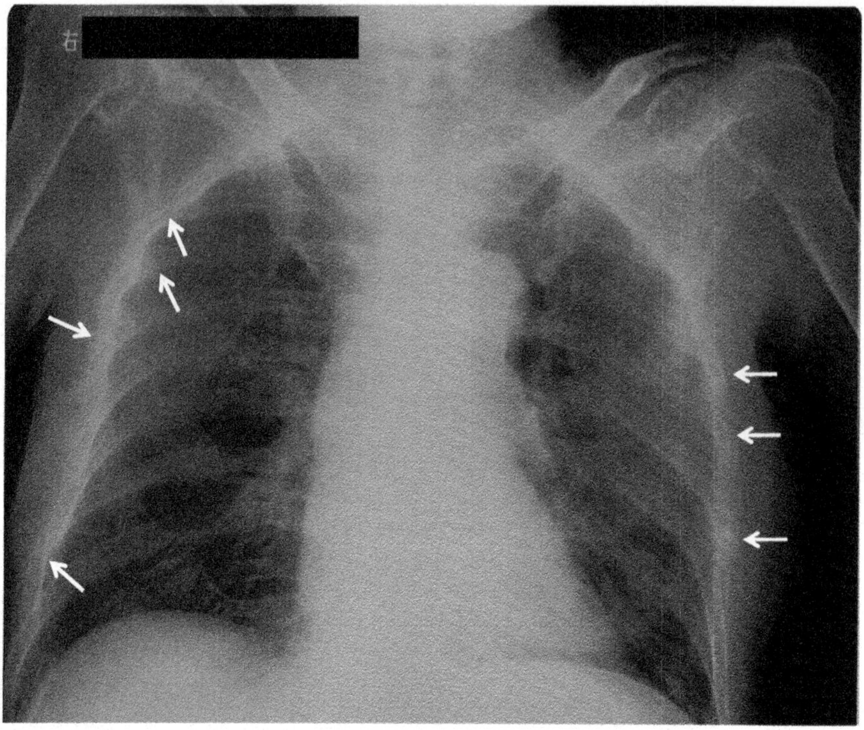

Fig. 2.2 Multiple pseudofractures (Looser's zones) in the ribs of a patient with *itai-itai* disease who was officially certified in 2008

2.4 The Mechanism of Cd Nephrotoxicity

Cd entering the body due to long-term exposure is excreted via the urine in minute amounts only. As a result, it predominantly accumulates in the liver and kidney cortex as the individual ages; accumulation in the liver continues up to approximately 80 years of age and reaches a plateau, whereas that in the kidney continues up to approximately 50 years of age and decreases with age due to increasing excretion through urine [11, 12]. The proximal tubules of the kidneys are the tissues most affected by Cd poisoning [1, 2, 11, 12]. In particular, Cd damages the S1 segment of the proximal tubules, inhibiting their ability to reabsorb glucose, amino acids, phosphates, bicarbonate, and low molecular weight proteins, among other substances. Accordingly, patients present with multiple disorders of proximal tubular reabsorptive function (known as Fanconi syndrome) [13]. As mechanisms underlying the reabsorption of proteins and other substances by the proximal tubular cells have been elucidated [14], the mechanism of Cd toxicity has become increasingly clear [5, 15]. Cd complexed to metallothionein (Cd-MT) in the bloodstream, which is released from the liver, is a low molecular weight molecule (molecular weight, $\leq$7000) and thus is freely filtered through the glomerulus. Filtered Cd-MT binds to receptors (e.g., megalin, cubulin) in the proximal tubules (S1, segment 1) before being transported into the cells via endocytosis, where it forms endosomes. Megalin separates from Cd-MT within the endosomes and is re-transported into the cell membrane. Endosomes containing Cd-MT separated from megalin then fuse with lysosomes. Metallothionein breaks down into amino acids through the action of hydrolases in lysosomes, and free Cd is then transported into the cytoplasm. Divalent metal transporter 1, a transporter of divalent metal ions (Cu^{2+}, Fe^{2+}, Zn^{2+}), is believed to be involved in the transportation of Cd into the cytoplasm. The stage during which free Cd shifts from the lysosomes to the cytoplasm is considered to be important in the cellular damage caused by Cd as this produces reactive oxygen species, the toxicity of which affects a variety of intracellular functions. Among these, damage to the Na^+ pump (Na^+/K^+-ATPase) has a major impact on basic cell functions and inhibits Na^+-dependent reabsorption of a variety of substances.

2.5 Pathophysiology of Cd Nephropathy

The developing mechanisms underlying proximal tubular function disturbances in Cd nephropathy are of great interest. We investigated the cross-sectional assessment of proximal tubular function in residents of the Jinzu River basin [16, 21]. One hundred forty-five residents of the Cd-contaminated Jinzu River basin who showed urinary β_2-MG ≥ 1 mg/g Cr (patient group) (age, 56–71 years; females, 78; mean age, 62.6 years; males, 67; mean age, 63.3 years) were divided into the following five groups based on fractional excretion of β_2-MG ($FE_{\beta 2\text{-}MG}$): <1%, ≥1% and <3%, ≥3% and <10%, ≥10% and <30%, and ≥30%. For comparison, 73 residents in a non-contaminated area (control group) (age, 56–71 years; females, 50; mean age, 63.5 years; males, 23; mean age, 64.9 years) were also examined. $FE_{\beta 2\text{-}MG}$ was investigated only in individuals with a urinary pH of ≥5.80, because β_2-MG is unstable in acid urine. Fractional excretions of urate (FE_{UA}), calcium (FE_{Ca}), sodium (FE_{Na}), and chloride (FE_{Cl}) were significantly higher in the patient group than in the control group, even in the $FE_{\beta 2\text{-}MG}$ <1% group in both males and females, indicating that FE_{UA}, FE_{Ca}, FE_{Na}, and FE_{Cl} increased from an extremely early stage of the disease. Comparisons of disease severity between the five $FE_{\beta 2\text{-}MG}$ groups indicated that FE_{UA} and the urinary excretion of protein and glucose increased with disease severity. On the other hand, investigation of tubular phosphorus reabsorption rates (%TRP), Cr clearance (Ccr), FE_{Ca}, and FE_{Na} demonstrated that %TRP and Ccr were at their lowest levels and FE_{Ca} and FE_{Na} levels were at their highest in the most severe group of ≥30%, while no significant differences were observed between the three groups of <1%, ≥1% and <3%, and ≥3% and <10%. Analysis of individual correlations with $FE_{\beta 2\text{-}MG}$ indicated strong correlations between FE_{UA} and the urinary excretion of protein and glucose, with correlation coefficients of at least 0.7. On the other hand, no correlations between %TRP, Ccr, FE_{Ca}, or FE_{Na} were observed, with correlation coefficients of <0.5. In other words, it is believed that the mechanisms underlying the reabsorption of uric acid, protein, and glucose are similar to that of β_2-MG, whereas those of phosphorus, calcium, sodium, and chloride differ from that of β_2-MG, which may indicate that the reabsorption of these elements is regulated by a variety of hormones and other substances.

2.6 Long-Term Clinical Course and Prognosis in Cd Nephropathy

2.6.1 *An 11-Year Follow-Up Study*

Multiple outbreaks of Cd nephropathy, with increased levels of urinary β_2-MG, have been recorded in residents of the Jinzu River basin [17]. An additional survey of the same population conducted 11 years later indicated the progression of urinary β_2-MG and glucose excretion levels [18, 19]. The results indicated significant increases in urinary β_2-MG and glucose in both males and females and that these increases were higher in residents of the Jinzu River basin than those of residents of the Ida or Kumano River basin (control group), indicating progression of proximal tubular dysfunction. A follow-up survey (1994–1995) conducted during a project to replace the contaminated rice paddy soil indicated that Cd concentrations in rice grown in private rice paddies by survey participants, who lived in areas where replacement had been completed, had significantly decreased. However, urinary Cd concentrations remained higher in residents of the Jinzu River basin than those of the control group, indicating that proximal tubular dysfunction caused by the long-term effects of Cd accumulation in the kidneys in high concentrations had been maintained or had progressed and worsened.

2.6.2 *A 27-Year Observational Study*

In 2011–2012, a 27-year observational study on 17 female Cd nephropathy patients was carried out to elucidate the progression of renal tubular dysfunction and erythropoiesis [20, 21]. The subjects were divided into the three groups based on urinary excretion of β_2-MG at the initial examination in 1985, as follows: $\geq$10 mg/g Cr (severe group; Cases 1–5; age, 83–92 years; mean age, 86.0 years in 2011–2012), <10 mg/g Cr and $\geq$5 mg/g Cr (moderate group; Cases 6–11; age, 82–91 years; mean age, 85.3 years), and <5 mg/g Cr (mild group; Cases 12–17; age, 82–90 years; mean age, 85.3 years) (Fig. 2.3). The 27-year changes in $FE_{\beta 2\text{-}MG}$ (Fig. 2.3a), maximal tubular reabsorption of phosphate corrected for the GFR (TmP/GFR) (Fig. 2.3b), serum Cr (Fig. 2.3c), and hemoglobin concentration (Fig. 2.3d) in the three groups were evaluated. In the severe group with urinary β_2-MG levels of $\geq$10 mg/g Cr, increased $FE_{\beta 2\text{-}MG}$ (Fig. 2.3a(A)), decreased TmP/GFR (Fig. 2.3b(A)), and increased serum Cr levels indicate markedly decreased GFR (Fig. 2.3c(A)). In addition, the onset of renal

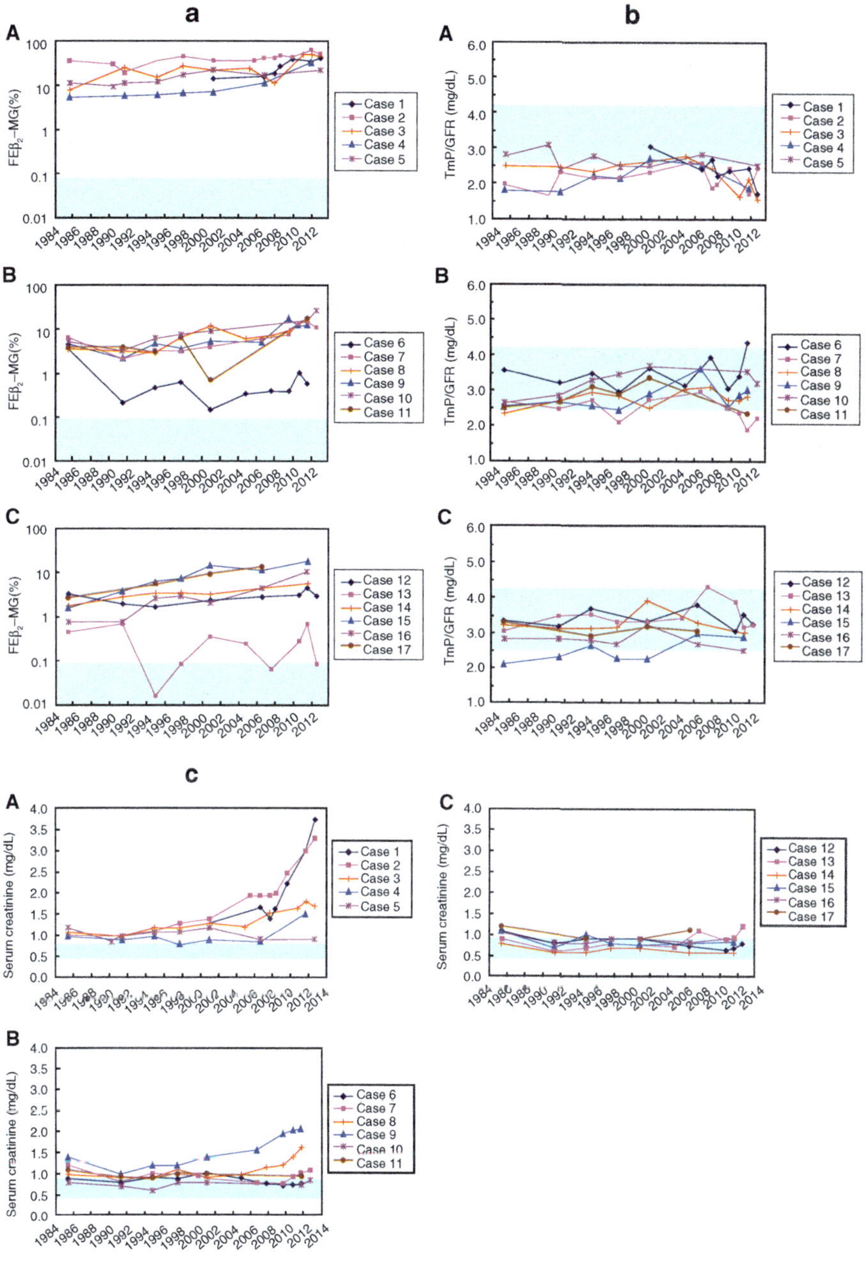

Fig. 2.3 27-year changes in (**a**) fractional excretion of β_2-microglobulin (β_2-MG), (**b**) TmP/GFR, (**c**) serum creatinine (Cr), (**d**) hemoglobin (Hb), (**e**) blood cadmium, and (**f**) urine cadmium in three groups of patients with Cd nephropathy as divided by urinary β_2-MG at the initial examination in 1985. (*A*) Severe group, urinary β_2-MG $\geq$10 mg/g Cr (Cases 1–5); (*B*) moderate group, urinary β_2-MG <10 mg/g Cr and $\geq$5 mg/g Cr (Cases 6–11); and (*C*) mild group, urinary β_2-MG <5 mg/g Cr (Cases 12–17). The shadow area represents the normal range

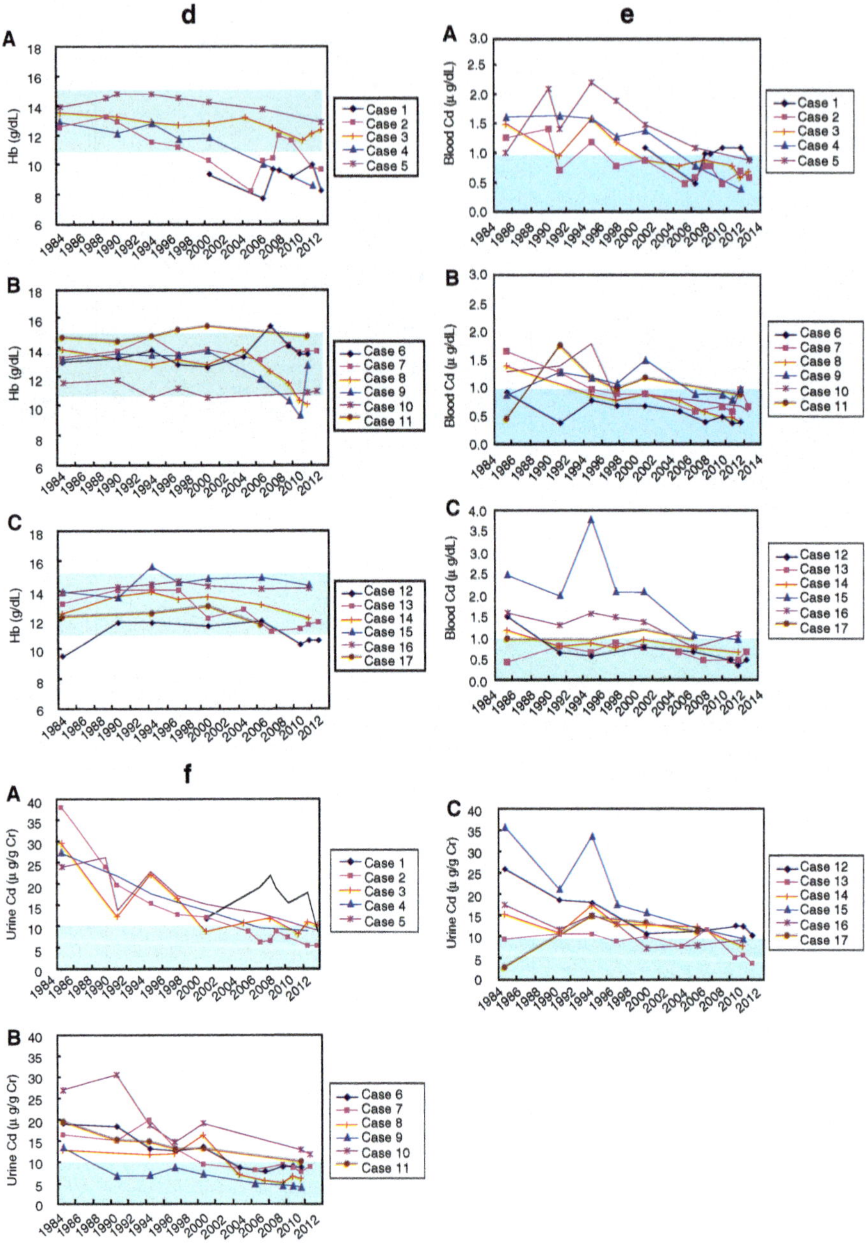

Fig. 2.3 (continued)

anemia in association with decreases in GFR below 30 mL/min was evident in the three patients (Cases 1, 2, and 4) (Fig. 2.3d(A)). The author has treated two patients (Cases 1 and 2) with erythropoietin (darbepoetin alfa), and an increase in hemoglobin concentration was seen as shown in Fig. 2.3d(A). Thus, erythropoiesis-stimulating hormone (recombinant human erythropoietin and darbepoetin alfa) successfully eliminated the anemia and the need for red cell transfusions, resulting in marked clinical improvement in patients with Cd nephropathy. One patient (Case 4) developed osteomalacia and was officially certified as having *itai-itai* disease in 2015. In the moderate groups with urinary β_2-MG of <10 mg/g Cr and ≥5 mg/g Cr, and in the mild groups with <5 mg/g Cr, $FE_{\beta 2\text{-MG}}$ levels increased (Fig. 2.3a(B, C)). However, the investigation of TmP/GFR (Fig. 2.3b) and GFR progressions indicating serum Cr levels (Fig. 2.3c) revealed differences between the three groups. Despite the clear tendency of decreases in the severe group as mentioned above, similar degrees of change were not observed in the moderate (Fig. 2.3b(B) and c(B)) and mild (Fig. 2.3b(C) and c(C)) groups. Therefore, cases of Cd nephropathy with urinary β_2-MG levels of ≥10 mg/g Cr develop progressive chronic renal failure, renal anemia, and metabolic bone disease and have a poor prognosis. These results indicated that urinary excretion level of β_2-MG predicts adverse outcome in Cd nephropathy. The 27-year changes in Cd in blood (Fig. 2.3e) and in urine (Fig. 2.3f) were also evaluated. Geometric means of Cd in the blood at the initial examination in 1985 and the follow-up in 2011–2012 were 1.28 μg/dL and 0.67 μg/dL, respectively, in the severe group ($p = 0.07$, Wilcoxon signed-rank test) (Fig. 2.3e(A)); 1.03 μg/dL and 0.64 μg/dL, respectively, in the moderate group ($p = 0.16$) (Fig. 2.3e(B)); and 1.22 μg/dL and 0.84 μg/dL, respectively, in the mild group ($p = 0.13$) (Fig. 2.3e(C)). The tendency of decreases was noted in the three groups, but the differences were not significant ($p \geq 0.05$). Geometric means of Cd in urine at the initial examination in 1985 and the follow-up in 2011–2012 were 24.6 μg/g Cr and 8.5 μg/g Cr, respectively, in the severe group ($p = 0.04$, Wilcoxon signed-rank test) (Fig. 2.3f(A)); 17.6 μg/g Cr and 8.1 μg/g Cr, respectively, in the moderate group ($p = 0.02$) (Fig. 2.3f(B)); and 14.3 μg/g Cr and 9.1 μg/g Cr, respectively, in the mild group ($p = 0.24$) (Fig. 2.3f(C)). Significant decreases were noted in the severe and moderate groups. Our longitudinal studies from 1983 to 2002 [4] showed a marked decline in the Cd levels of rice grains as a result of soil replacement in the contaminated paddy fields: these results may partly explain the decreased blood and urine Cd levels in the 17 cases.

2.7 FGF-23 in Cd Nephropathy and Patients with *Itai-Itai* Disease

FGF-23 is a hormone that is produced by osteocytes and regulates phosphate and vitamin D metabolism through binding to the Klotho-FGF receptor complex in proximal tubules [7, 22]. Thus, FGF-23 secreted by osteocytes suppresses phosphorus reabsorption by decreasing the expression of type IIa and IIc sodium-phosphate cotransporters on the apical membrane of proximal tubular cells. Moreover, it also inhibits 1α-hydroxylase activity, whereas it stimulates 24-hydroxylase activity, thus leading to decreased 1, 25-dihydroxyvitamin D serum levels. Excessive actions of FGF-23 cause several kinds of hypophosphatemic osteomalacia [7]. *Itai-itai* disease is a Cd-induced renal tubular osteomalacia. Hypophosphatemia is common in *itai-itai* disease and Cd nephropathy [23, 24]. Therefore, we investigated whether FGF-23 is involved in the etiology of hypophosphatemia in *itai-itai* disease and Cd nephropathy in 2014–2015.

Four patients with *itai-itai* disease (females, two; males, two; age, 80–94 years; mean age, 87.7 years), 47 subjects (females, 35; males, 12; age, 67–94 years; mean age, 82.6 years) living in the Cd-contaminated Jinzu River basin, and 7 subjects (females, five; males, two; age, 67–90 years; mean age, 80.0 years) living in non-contaminated districts in Toyama (references) were studied. Serum concentrations of FGF-23 were determined using a sandwich enzyme-linked immunosorbent assay. The values for serum FGF-23 in the four patients with *itai-itai* disease ranged from 22.9 to 719.2 pg/mL with a geometric mean of 124.8 pg/mL. FGF-23 in the 47 subjects ranged from 10.6 to 155.7 pg/mL with a geometric mean of 35.7 pg/mL, while that in the 7 references ranged from 36.0 to 82.1 pg/mL with a geometric mean of 53.2 pg/mL. The 47 subjects from the Cd-contaminated Jinzu River basin were divided into 4 groups according to the severity of proximal tubular dysfunction based on their urinary excretion of β_2-MG levels as follows: $\geq$100 mg/g Cr (severe group; 9 individuals; females, 8; male, 1; age, 77–94 years; mean age, 86.5 years), <100 mg/g Cr and $\geq$5 mg/g Cr (moderate group; 14 individuals; females, 8; males, 6; age, 70–99 years; mean age, 85.0 years), <5 mg/g Cr and $\geq$1 mg/g Cr (mild group; 11 individuals; females, 7; males, 4; age, 71–86 years; mean age, 82.6 years), and <1 mg/g Cr (normal group; 13 individuals; females, 12; male, 1; age, 67–86 years; mean age, 77.2 years). Figure 2.4a shows the distribution of serum FGF-23 in these four groups. Geometric means of serum FGF-23 in the four groups (30.8 pg/mL in the severe group, 32.9 pg/mL in the moderate group, 37.5 pg/mL in the mild group, and 41.5 pg/mL in the normal group, respectively) tended to decrease with increasing severity of proximal tubular dysfunction, but the difference was not significant ($p = 0.327$, the Kruskal–Wallis test). Figure 2.4b shows the relationship between serum levels of FGF-23 and phosphorus. The serum level of FGF-23 was significantly and positively correlated with serum phosphorus level at the level of $r = 0.624$ ($p < 0.0001$) in the 47 subjects and 3 patients with *itai-itai* disease, with the exception of 1 patient which is indicated by an arrow in Fig. 2.4b. These results indicated that increasing serum phosphorus concentration induces secretion of FGF-23. Figure 2.4c shows the relationship between serum levels of FGF-23 and TmP/GFR. In 19 subjects with TmP/GFR below 2.5 mg/dL, except the 3 cases with extremely elevated levels of FGF-23 above 100 pg/mL, the levels of serum FGF-23 appeared to be positively related to TmP/GFRs, with a correlation coefficient of 0.467 ($p = 0.044$). Figure 2.4d shows the relationship between serum levels of FGF-23 and cystatin-C-based estimated GFR (eGFRcys). The two cases with extremely elevated levels of FGF-23 above 100 pg/mL had extremely decreased levels of eGFRcys: 4.1 and 7.4 mL/min/1.73 m^2, respectively. These results indicated that decreased GFRs induce increasing serum phosphorus, which stimulates greater secretion of FGF-23 in *itai-itai* disease and Cd nephropathy. One patient

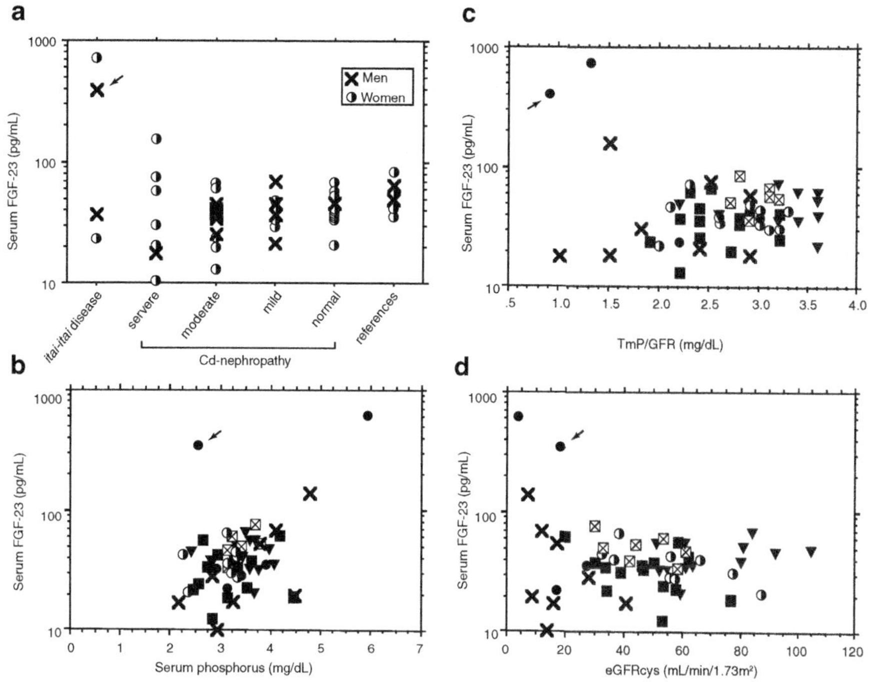

Fig. 2.4 (**a**) Serum levels of fibroblast growth factor (FGF) 23 in six groups of subjects: *itai-itai* disease patients, four groups with Cd nephropathy, and references. (**b**) The relationships between serum FGF-23 and serum phosphorus, (**c**) between serum FGF-23 and TmP/GFR, and (**d**) between serum FGF-23 and eGFRcys in six groups of subjects. ● *itai-itai* disease patients, *X* severe stage of Cd nephropathy, ■ moderate stage of Cd nephropathy, ◑ mild stage of Cd nephropathy, ▼ normal stage of Cd nephropathy, and ⊠ references. (Aoshima K. et al., unpublished)

with *itai-itai* disease had inappropriately high serum FGF-23 despite relatively low serum phosphorus and is indicated with an arrow in Fig. 2.4b. These findings may lead to the diagnosis of FGF-related osteomalacia in this patient; further examinations are needed to determine the cause of the increased serum level of FGF-23. In summary, FGF-23 is not involved in the pathogenesis of hypophosphatemia in *itai-itai* disease and Cd nephropathy. Renal phosphate wasting due to Cd-induced proximal tubular dysfunction is the primary mechanism of hypophosphatemia in *itai-itai* disease, which is known as Cd-induced renal tubular osteomalacia [23, 24].

2.8 Conclusions

The present chapter provides an overview of results obtained from studies on the pathophysiology and progression of Cd nephropathy among residents of the Cd-contaminated Jinzu River basin in Toyama Prefecture. In this area, a marked decline in the Cd levels of rice grains has been achieved due to the replacement of contaminated paddy field soil, which was conducted over 33 years, from 1979 until completion in 2011. However, Cd nephropathy is still prevalent, progressive, and irreversible in residents who have been exposed to Cd for a long time. New cases of *itai-itai* disease are still being detected as residents age (typically over 80 years old) and are diagnosed in both males and females. Therefore, continued screening and long-term monitoring of Cd nephropathy are essential for residents of this area, both to clarify the natural history of chronic Cd poisoning and for an early diagnosis to mitigate the adverse effects as much as possible.

Acknowledgments The research results covered in the present paper include the results of the "Study on the health effects of environmental exposure to cadmium in the residents of cadmium-contaminated Jinzu River basin" (principal investigator: Keiko Aoshima), which were parts of the 2012 and 2016 Studies of the Health Effects of Heavy Metals organized by the Ministry of the Environment, Japan.

Conflict of Interest The author has no conflict of interest to declare.

References

1. Friberg L, Piscator M, Nordberg GF, Kjellstrom T. Health effects of cadmium in the general environment in Japan. In: Cadmium in the environment. 2nd ed. Cleveland: Chemical Rubber Press; 1974. p. 137–227.
2. World Health Organization (WHO). Environmental health criteria 134. Cadmium. Geneva: WHO; 1992.
3. Nogawa K. *Itai-itai* disease and follow-up studies. In: Nriagu JO, editor. Cadmium in the environment, part II. New York: John Wiley & Sons; 1981. p. 643–89.
4. Aoshima K. *Itai-itai* disease: renal tubular osteomalacia induced by environmental exposure to cadmium–historical review and perspectives. Soil Sci Plant Nutr. 2016;62:319–26.
5. Aoshima K. *Itai-itai* disease: cadmium-induced renal tubular osteomalacia–current situations and future perspectives. Jpn J Hyg. 2012;67:455–63. (in Japanese with English abstract)
6. Aoshima K. *Itai-itai* disease: lessons from the investigations of environmental epidemiology conducted in the 1970's, with special reference to the studies of the Toyama Institute of Health. Jpn J Hyg. 2017;72:149–58. (in Japanese with English abstract)
7. Fukumoto S. Physiological regulation and disorders of phosphate metabolism–pivotal role of fibroblast growth factor 23. Int Med. 2008;47:337–43.
8. Kinyobinews (January 24, 2014). http://www.kinyobi.co.jp/kinyobinews/2014/01/24. Accessed 17 Sept 2017.
9. Nakagawa S. A study of osteomalacia in Toyama prefecture (so-called *itai-itai* disease). Ser Department Science, Kanazawa Univ Med Sch. 1960;56:1–51. (in Japanese with English summary)
10. Ministry of the Environment. Report on cadmium contaminated area resident health impact study committee; 2009 (in Japanese)
11. Wedeen RP, De Broe ME. Nephrotoxic metals. In: Davison AM, Cameron JS, Grunfeld JP, Ponticelli C, Ritz E, Winearls CG, et al., editors. Oxford textbook of clinical nephrology. 3rd ed. Oxford: Oxford University Press; 2005. p. 1075–89.

12. Nordberg GF, Nogawa K, Nordberg M, Friberg L. Cadmium. In: Nordberg GF, Fowler BA, Nordberg M, Friberg L, editors. Handbook on the toxicology of metals. Elsevier: London; 2007. p. 445–86.
13. Brewer ED. The Fanconi syndrome. In: Gonick HC, Buckalew Jr VM, editors. Renal tubular disorders. New York: Marcel Dekker; 1985. p. 475–544.
14. Van't Hoff WG. Fanconi syndrome. In: Davison AM, Cameron JS, Grunfeld JP, Ponticelli C, Ritz E, Winearls CG, et al., editors. Oxford textbook of clinical nephrology. 3rd ed. Oxford: Oxford University Press; 2005. p. 961–73.
15. Thevenod F. Nephrotoxicity and the proximal tubule. Insight from cadmium. Nephron Physiol. 2003;93:87–93.
16. Aoshima K, Kawanishi Y, Fan J, Cai Y, Katoh T, Teranishi H, et al. Cross-sectional assessment of renal function in the inhabitants of a cadmium-polluted area. Ann Clin Lab Sci. 1995;25:493–503.
17. Aoshima K. Epidemiology of renal tubular dysfunction in the inhabitants of a cadmium-polluted area in the Jinzu River basin in Toyama prefecture. Tohoku J Exp Med. 1987;152:151–72.
18. Fan J, Aoshima K, Katoh T, Teranishi H, Kasuya M. A follow-up study on renal tubular dysfunction in women living in the cadmium-polluted Jinzu River basin in Toyama, Japan: part 1. Changes in the level of exposure to cadmium after soil replacement of polluted paddy fields and the related effects on the prognosis of renal tubular dysfunction. Jpn J Hyg. 1998;53:545–57. (in Japanese with English abstract)
19. Cai Y, Aoshima K, Katoh T, Teranishi H, Kasuya M. Renal tubular dysfunction in male inhabitants of a cadmium-polluted area in Toyama, Japan: an eleven-year follow-up study. J Epidemiol. 2001;11:180–9.
20. Aoshima K, Katoh T, Kasuya M. Renal tubular function and bone metabolism in inhabitants of the cadmium-polluted Jinzu river basin in Toyama Prefecture, Japan: a 27-year follow-up study of 17 women with β_2-microglobulinuria. The 2012 Ministry of the Environment commissioned study report: General research on the relation between *itai-itai* disease and chronic cadmium poisoning; 2013. p. 47–63. (in Japanese with English summary)
21. Aoshima K, Ogura Y. *Itai-itai* disease and kidneys: epidemiology, pathophysiology and clinical course in multiple proximal tubular dysfunction induced by environmental exposure to cadmium. Kidney Metab Bone Dis. 2016;29:332–40. (in Japanese with English summary)
22. Shimada T, Hasegawa H, Yamazaki Y, Muto T, Hino R, Takeuchi Y, et al. FGF-23 is a potent regulator of vitamin D metabolism and phosphate homeostasis. J Bone Miner Res. 2004;19:429–35.
23. Aoshima K, Iwata K, Kasuya M. Environmental exposure to cadmium and effects on human health. Part 2. Bone and mineral metabolism in inhabitants of the cadmium-polluted Jinzu River basin in Toyama prefecture. Jpn J Hyg. 1988;43:864–71. (in Japanese with English abstract)
24. Aoshima K, Katoh T, Teranishi H, Horiguchi H, Kasuya M. Abnormalities of calcium, phosphorus and vitamin D metabolism with proximal renal tubular dysfunction in subjects environmentally exposed to cadmium. Jpn J Hyg. 1993;47:1009–20. (in Japanese with English abstract)

Chapter 3
Novel Pathological Study of Cadmium Nephropathy of *Itai-itai* Disease

Johji Imura, Koichi Tsuneyama, and Yoshihiko Ueda

Abstract *Itai-itai* disease, the most severe form of chronic cadmium (Cd) poisoning, is caused by prolonged ingestion of Cd. It developed in numerous inhabitants of the Jinzu River basin in Toyama Prefecture, Japan, an area severely polluted by Cd. The target organ of Cd toxicity is the kidney, and in particular, it is known that this disease is characterized mainly by renal tubule disorder and subsequent renal osteomalacia, resulting from proximal tubular injury. This chapter is to clarify the pathogenesis of tubulopathy caused by Cd exposure in *itai-itai* disease kidney by novel pathological study.

Paraffin-embedded sections of renal tissues were obtained from a total of 98 autopsied cases of *itai-itai* disease in the University of Toyama. Thirty-six autopsy cases of other disease without Cd exposure as control group were selected. The extent of hyalinized nephron and interstitial fibrosis was assessed by Masson trichrome staining. In the immunohistochemical study, each antibody used was liver fatty acid-binding protein (LFABP) of the liver recognized to the proximal tubule, CD34 for perivascular capillaries, and D2-40 for lymphatic vessels. Immunohistovisualization was done by automated immunostainer. In morphometrical analysis, immunoreactive and staining areas were evaluated by morphometrical analysis.

The marked atrophy of proximal tubules was exhibited in the patient of *itai-itai* disease. Immunohistochemical study using LFABP revealed that the proximal tubular region was significantly reduced. On the other hand, with regard to nephron

J. Imura (✉)
Department of Diagnostic Pathology, Graduate School of Medicine and Pharmaceutical Sciences, University of Toyama, Toyama, Toyama, Japan
e-mail: imura@med.u-toyama.ac.jp

K. Tsuneyama
Department of Molecular and Environmental Pathology, Institute of Health Biosciences, The University of Tokushima Graduate School, Tokushima, Tokushima, Japan

Y. Ueda
Department of Diagnostic Pathology, Saitama Medical Center, Dokkyo University School of Medicine, Koshigaya, Saitama, Japan

S. Himeno, K. Aoshima (eds.), *Cadmium Toxicity*, Current Topics in Environmental Health and Preventive Medicine,
https://doi.org/10.1007/978-981-13-3630-0_3

disorder, the global sclerosis of glomerulus was observed in the superficial nephron which is present in the outer layer of renal cortex. Interstitial fibrosis was recognized to repair the disordered region. Altitude disappearance of peritubular vascular network and dilation of the lymph duct was confirmed by immunohistochemical studies with CD34 and D2-40 antibody.

In this study we were able to elucidate the mechanism of various pathogenesis in *itai-itai* disease. These findings are important discovery to explain the clinical evidence of the polyuria in *itai-itai* disease. It was objectively reconfirmed that the main injured site in the kidney of this disease is mainly the proximal renal tubule, as instructed in previous histological studies. As a new finding, there is also a disability of nephron which has not attracted attention so far. Namely, it has been reported that the outer nephron was correlated with osmotic adjustment of urine from the physiologic studies. Therefore, it is supposed that the polyuria might appear by disturbance of the urine volume adjustment in the patients of *itai-itai* disease that these nephrons were impaired. Although the mechanism about the decrease of peritubular capillaries and the dilation of lymphatic vessels is unclear, these findings may explain the mechanism of tubular impairment. An additional study is needed. In conclusion, although there have been few previous reports on the disorder of nephrons in Cd nephropathy, this study demonstrated disorders of superficial nephrons were more specific for *itai-itai* disease than for other renal diseases. Furthermore, the abnormalities of peritubular capillaries and lymphatic vessels in conjunction with the superficial nephron may become the key to elucidate the mechanism of this disease.

Keywords *Itai-itai* disease · Pathological study · Autopsy · Renal lesion Immunohistochemistry

3.1 Introduction

Itai-itai disease, the most severe form of chronic cadmium (Cd) poisoning, is caused by prolonged ingestion of Cd. *Itai-itai* disease developed in numerous inhabitants of the Jinzu River basin in Toyama Prefecture, Japan, an area severely polluted by Cd. The disease is mainly characterized by renal tubular disorder (tubulopathy) and subsequent renal osteomalacia. The pathogenesis of renal osteomalacia has been reported in previous studies [1–4]. This report aims to provide additional information about the pathogenesis of tubulopathy caused by Cd exposure in *itai-itai* disease kidney. Disorders of various organs may be caused by prolonged oral ingestion of heavy metals, such as Cd. The kidneys are the main target organs of Cd toxicity, with epithelial damage of the proximal tubules being particularly characteristic. The abnormalities of calcium and phosphorus metabolism, caused by tubular extinction and subsequent osteomalacia, are essential mechanisms in *itai-itai* disease. Here, we report that lesions in the outer layer nephrons, and changes in peritubular capillary networks associated with nephrons, are specifically found in *itai-itai* disease.

Previous studies, based on morphological observations, highlight the proximal tubule as the main locus of the lesion in *itai-itai* disease. However, there has not been enough discussion on the presence or absence of disturbance in the distal renal tubule. Furthermore, additional clarification is required to investigate changes exhibited by other renal components, such as nephrons typified by glomeruli, rather than focusing solely on tubules. Further study was carried out with the aim to address this knowledge gap.

3.2 Materials and Methods

To study histopathology and immunohistochemistry, formalin-fixed and paraffin-embedded sections were used. Samples were obtained from *itai-itai* disease patients of 98 cases autopsied in Toyama University. The control group used kidneys from 36 female patients, of the same age, who died from diseases other than *itai-itai* disease, such as malignancy of several organs, pneumonia, cerebral bleeding, myocardial infarction, etc. The Masson trichrome method was used to observe interstitial fibrosis and hyalinized nephrons. Immunohistochemically, anti-liver fatty acid-binding protein (LFABP) antibody was used as a marker of the proximal tubular epithelium. CD34 and D2-40 were used to identify the peritubular capillary and the lymphatic vessels, respectively. BenchMark GX® automated IHC/ISH slide staining system (Roche) was used to visualize antibody staining. Morphometrical analysis to evaluate degree of staining was calculated by WinRoof (Mitani, Co.).

3.3 Results

3.3.1 The General Pathological Findings in Itai-itai *Disease*

Firstly, the general pathological findings of renal lesions in the *itai-itai* disease kidney will be described. Macroscopically, the kidneys of *itai-itai* disease patients were markedly atrophic, weighing an average of 40 g (Fig. 3.1a). Renal atrophy was high in *itai-itai* disease-affected kidneys, and the kidney surface was relatively smooth or had a fine granular appearance. This has been described as "sandpaper-like appearance" (Fig. 3.1b). In contrast, the kidneys of patients with benign nephrosclerosis- and arteriosclerosis-related lesions had irregular surface. In *itai-itai* disease patients, on the cut surface, the renal cortex was particularly atrophic, while the medulla was preserved (Fig. 3.1c).

Next, the histopathological findings will be outlined. There were no primary or secondary proximal tubular lesions in the kidney of the control group. Nephropathic lesions in the *itai-itai* disease-affected kidney were mainly present in the proximal tubule (Fig. 3.2a). The proximal tubules were atrophic, desquamative, or

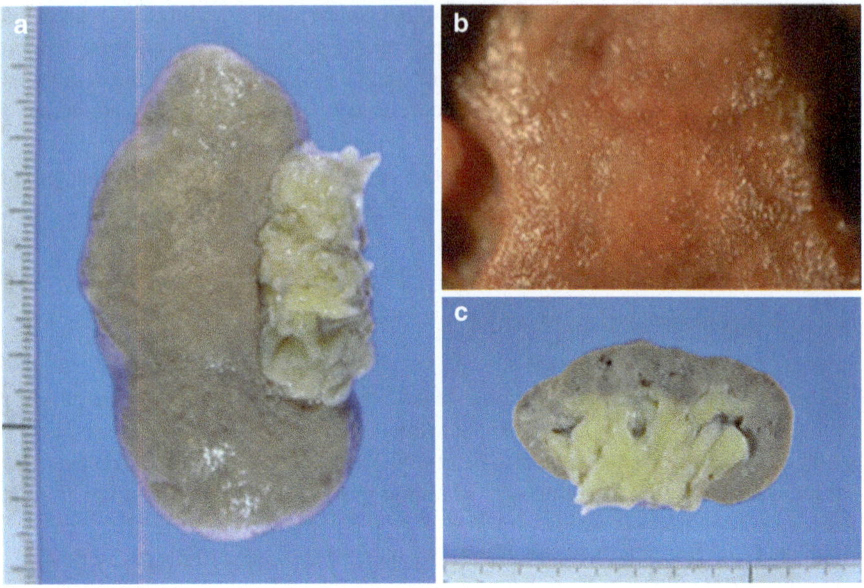

Fig. 3.1 Macroscopic findings of the kidney in *itai-itai* disease patient. Markedly atrophic kidney (**a**) revealed fine granular surface (**b**) and thinning renal cortex (**c**)

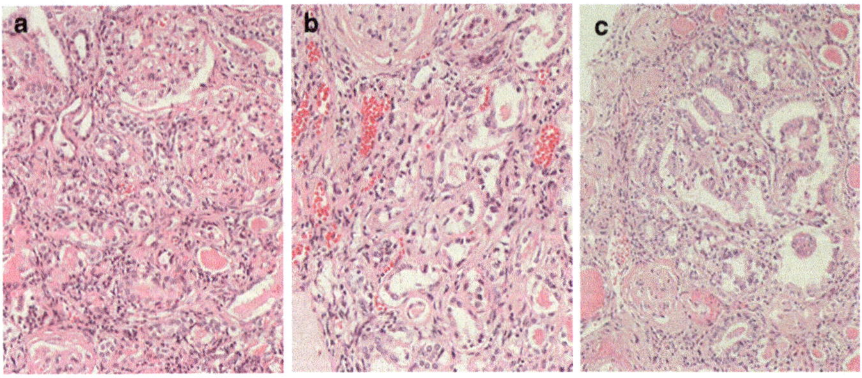

Fig. 3.2 Histopathological findings of the kidney in *itai-itai* disease patient. The proximal tubules showed marked atrophy (**a**), flattering, desquamation, and/or disappearance of tubular epithelium (**b**). The distal tubules showing the hyperplastic changes (**c**)

disappearing, and the tubular epithelium appeared flat (Fig. 3.2b). Conversely, the distal tubule showed rather hyperplastic changes; it is possible that this could be a result of efforts to compensate for the function of the disordered proximal renal tubule (Fig. 3.2c). In the secondary disappearance of the proximal tubules, interstitial fibrosis also progresses. Furthermore, the remaining renal tubules followed the progress of fibrosis and atrophy (Fig. 3.3a, b). Active inflammation, such as interstitial nephritis, in the stroma can be partly seen (Fig. 3.3c). As other findings, so-called thyroid-like appearance refers to findings similar to the colloidal follicle

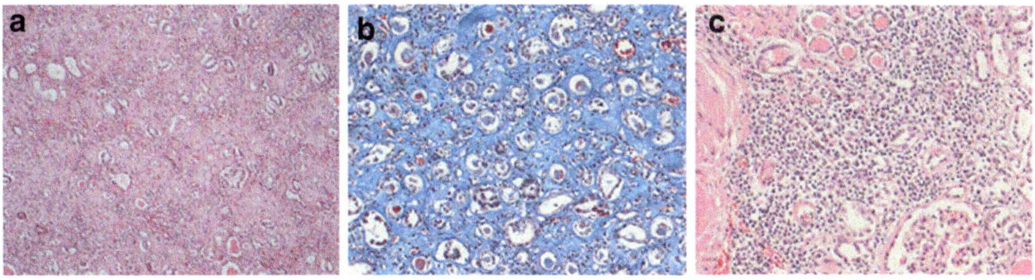

Fig. 3.3 Histopathological findings of the kidney in *itai-itai* disease patient. Peripheral fibrosis observed around the atrophic proximal renal tubule (**a**), Masson trichrome staining (**b**). Active inflammatory cell infiltration seen in the stroma (**c**)

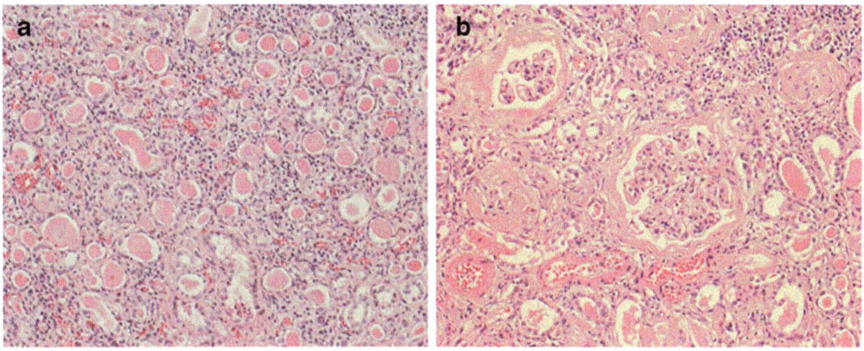

Fig. 3.4 Histopathological findings of the kidney in *itai-itai* disease patient. Dilated distal tubules containing amorphous material suggesting chronic pyelonephritis (**a**). The fibrosis seen in the Bowman's capsule (**b**)

of the thyroid gland caused by repeated pyelonephritis that are recognized in the many cases. Concentrated urine was retained in the lumen of distal tubules as pinkish amorphous materials (Fig. 3.4a). Interstitial nephritis of the renal parenchyma and peri-Bowman's fibrosis were also observed (Fig. 3.4b).The renal parenchyma exhibited a characteristic lesion as well as proximal tubular impairment. The nephron disorder is focused mainly on the glomeruli. Most glomeruli become extinct, and the lesion, so-called hyalinized glomerulus or global glomerular sclerosis, was present. The impaired glomerulus was present only in the outer layer of the renal cortex (Fig. 3.5a, b).

3.3.2 *The Additional Study of Renal Tubular and/or Nephron Disorder*

The control group samples exhibited positive reactions for LFABP in the proximal tubule area. In *itai-itai* disease patient kidneys, the positive area was significantly decreased compared to the control group (Fig. 3.6a–e).

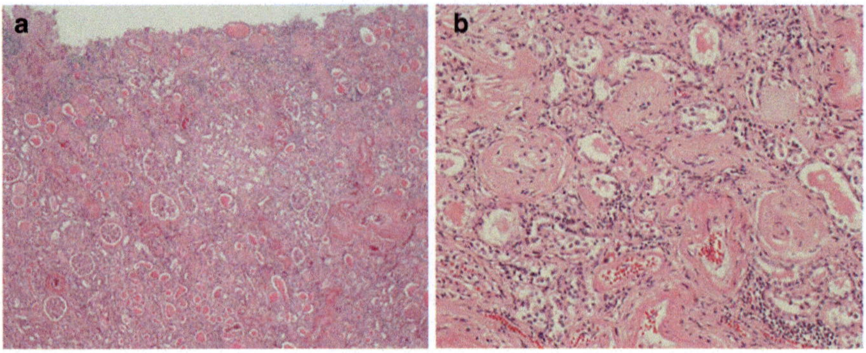

Fig. 3.5 Histopathological findings of the kidney in *itai-itai* disease patient. Superficial glomerulus of renal cortex showing hyalinization or global sclerosis (Fig. 3.5a, b)

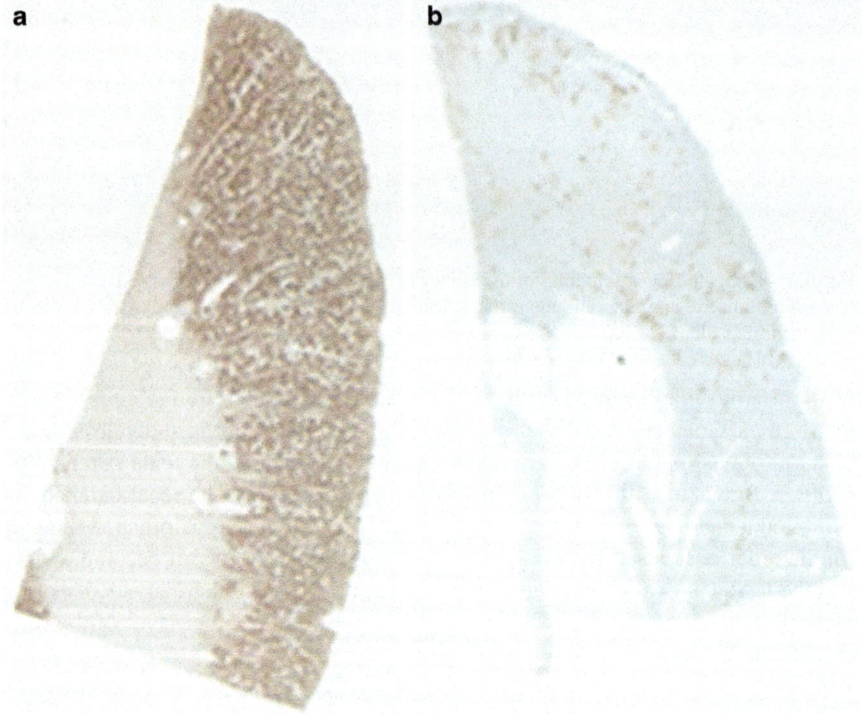

Fig. 3.6 Immunohistochemical findings of LFABP expression in the kidneys of control group (**a**, **b**) and *itai-itai* disease patient (**c**, **d**). The kidney of control group showed positive reaction in the preserved proximal tubules; *itai-itai* disease group revealed the decrease area of positive tubules. The comparison of immunoreactive area for LFABP immunity between control and *itai-itai* disease group (**e**)

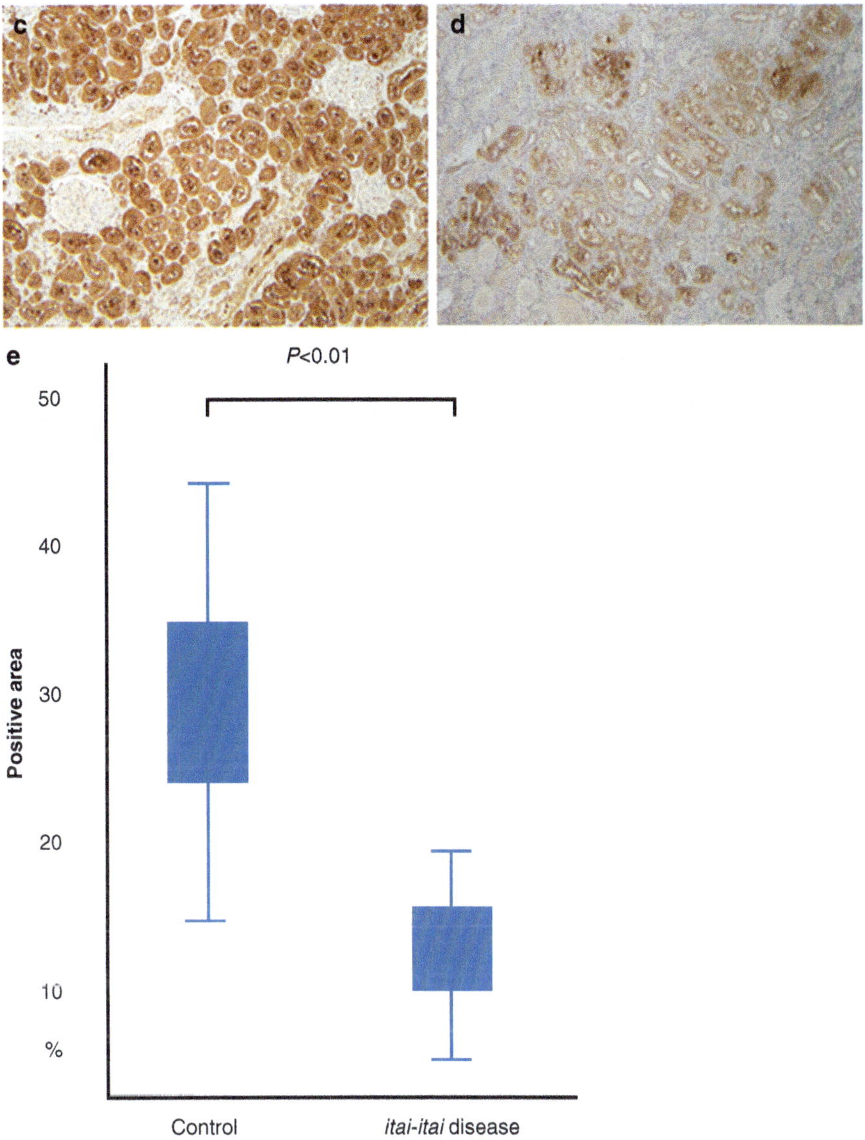

Fig. 3.6 (continued)

The hyalinized nephrons of *itai-itai* disease patients were mainly located in the superficial layer. In contrast, the lesions of the control group samples were mainly located in the deep layer. In order to objectively investigate which layer of the nephron was injured, the distance from the renal surface to the hyalinized nephron was measured. Comparing the *itai-itai* disease group to the control group, *itai-itai* disease-affected kidneys showed a shorter distance from the renal surface to the hyalinized nephron compared to the control group samples. This suggests that the injured nephrons are mainly in superficial layers of the renal cortex (Fig. 3.7a–c).

To examine interstitial fibrosis of the kidney, Masson trichrome staining method was used to evaluate the collagenous fibrotic area. In comparison to the control group, the area of fibrosis was significantly larger in the kidney of *itai-itai* disease patients (Fig. 3.8a–c).

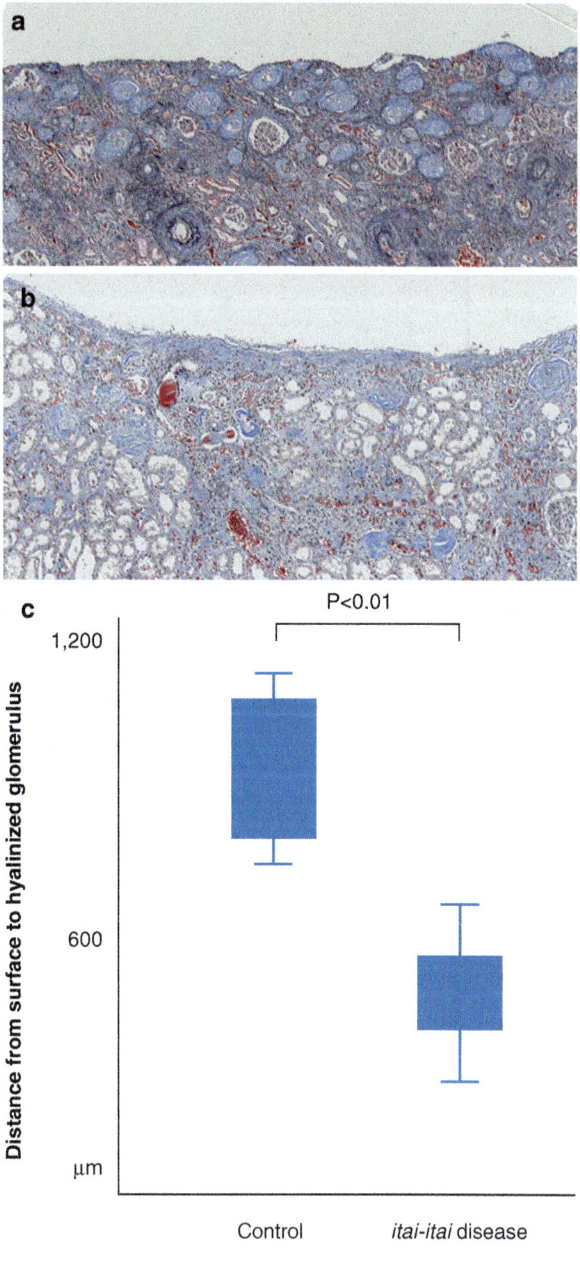

Fig. 3.7 The disturbance of nephrons. Hyalinized nephrons of *itai-itai* disease kidney (**a**) mainly located in the outer layer compared with the control group (**b**). The distance from the surface to the hyalinized glomerulus is predominantly short (**c**)

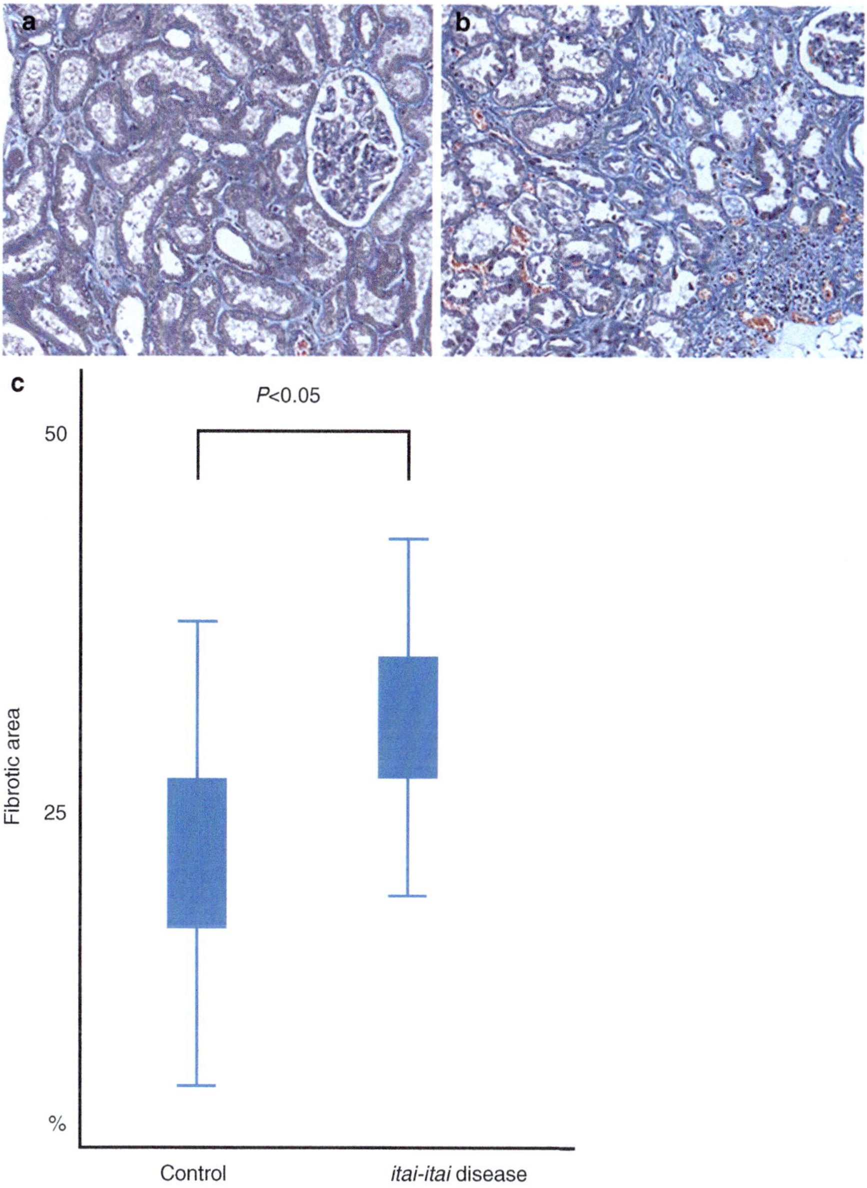

Fig. 3.8 Interstitial fibrosis in the peritubular area. The collagenous fibrosis observed in the peritubular areas by the Masson trichrome staining in *itai-itai* disease (**a**) and control group (**b**). The blue staining area of *itai-itai* disease kidney decreasing by the morphometrical analysis (**c**)

The capillary network around the renal tubule was immunohistochemically evaluated using CD34 antibody. Richly developed capillary networks were observed in the control group (Fig. 3.9a, b). In *itai-itai* disease patients, the peritubular capillary network decreased remarkably. In the immunohistochemical study using D2-40, lymphatic vessels were not observed in the normal kidneys; however a large number

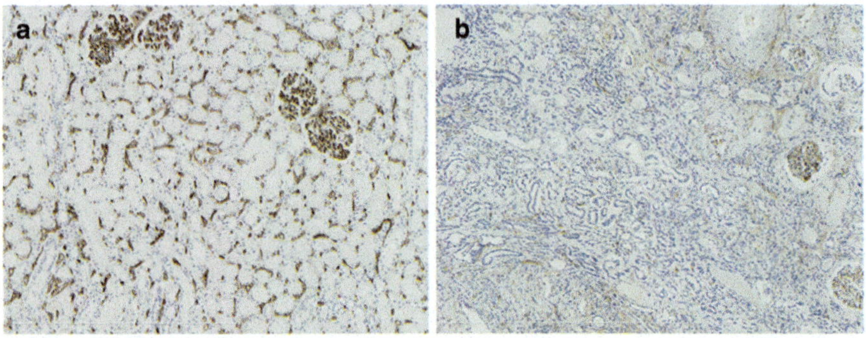

Fig. 3.9 Immunohistochemical findings of peritubular capillaries for CD34 in control (**a**) and *itai-itai* disease group (**b**). The markedly decreasing positive area in *itai-itai* disease comparing with control group

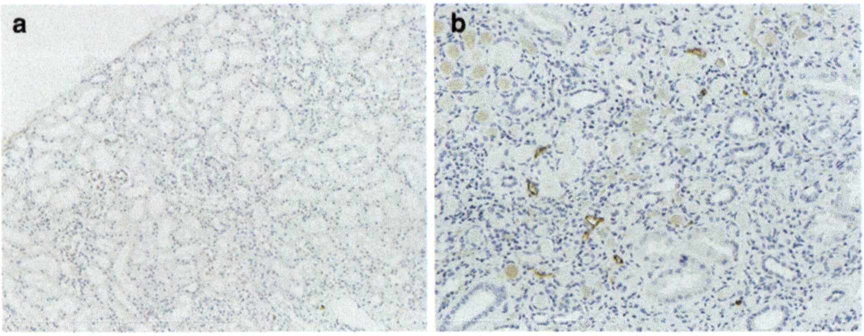

Fig. 3.10 Immunohistochemical findings of peritubular lymphatic vessels for D2-40 in control (**a**) and *itai-itai* disease group (**b**). Dilated lymphatic vessels observed in the *itai-itai* disease comparing with control group

of the dilated lymphatic vessels were identified in the kidneys of *itai-itai* disease patients (Fig. 3.10a, b).

3.4 Discussion

From results of the present study, the relationship between the pathological kidney findings and *itai-itai* disease patient symptoms is summarized in the Fig. 3.11. Proximal tubule disorder primarily occurs because of Cd exposure [5, 6]. In the present study, the marked atrophy of proximal tubules was exhibited in *itai-itai* disease patients, and the degree of atrophy was quantified immunohistochemically using LFABP. We added an immunohistochemical study of LFABP in order to objectively recognize the degree of injury of proximal tubule. Although no marker that accurately recognizes the proximal renal tubule existed so far, LFABP localizes

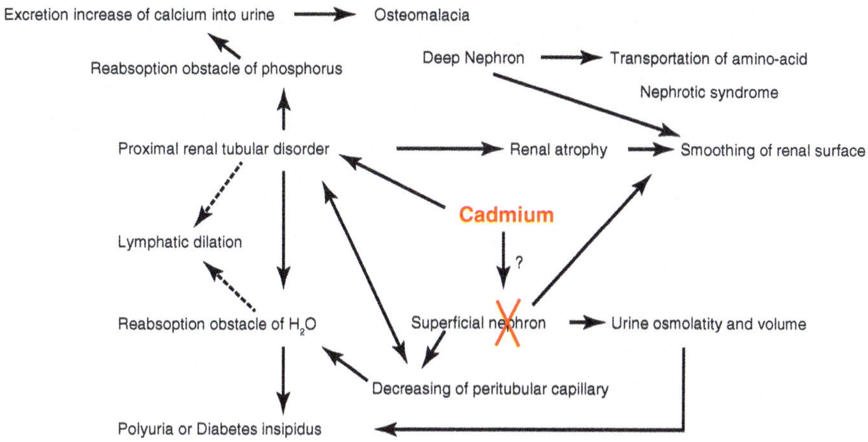

Fig. 3.11 Hypothesis of pathogenesis for the renal dysfunction in the *itai-itai* disease patients

in the proximal renal tubule [7] and appears in the urine when these sites are impaired; therefore, it came to be used in recently widely clinical tests. The reabsorption obstacle of phosphorus in the proximal tubule occurred, and then it has been thought that calcium flows out into the urine due to the osteomalacia. A new pathophysiological concept of *itai-itai* disease became clear from the present study. A nephron disorder was recognized in the outer layer of the renal cortex. These findings are an important discovery and help to explain the clinical evidence of polyuria in *itai-itai* disease [8]. Physiological studies report that outer nephrons correlate with osmotic adjustment of urine [9]. Therefore, it is proposed that polyuria might be explained by disturbance of urine volume in *itai-itai* disease patients where these nephrons were impaired. Although the degree of renal atrophy was high, the surface was smooth. This is linked with hyalinization occurring only in the superficial nephrons of the renal cortex. Furthermore, superficial nephrons regulate urinary osmotic pressure and urine volume. In contrast, the deep nephrons mainly transport amino acid and/or proteins [9]. The difference of both can explain the general symptoms of an *itai-itai* disease patient. An *itai-itai* disease patient characteristically has polyuria up to 4000 mL per day, but this rarely results in nephrotic syndrome [8].The patients feature polyuria, because the superficial nephrons are affected. The deep nephrons do not cause nephrotic syndrome because they are retained.

In addition, the capillary vessel network around the renal tubule is closely connected to the superficial nephrons. As the superficial nephrons hyalinization, capillary network function around the renal tubule may be impacted. As a result, reabsorption of the water does not occur and this results in polyuria. On the other hand, it is inferred that edema and lymphatic dilation appears in the kidney. We thought that the above is nephropathic pathophysiologic condition in *itai-itai* disease patients. While additional studies are needed to explain the mechanisms causing the decrease of peritubular capillaries and the dilation of lymphatic vessels, our findings may explain the mechanism of tubular impairment.

In conclusion, although there have been previous reports on the disorder of nephrons in Cd nephropathy, this study demonstrated that disorders of superficial nephrons were more specific for *itai-itai* disease than for other renal diseases. Furthermore, the abnormalities of peritubular capillaries and lymphatic vessels, in conjunction with the superficial nephrons, may become central to elucidating the mechanism of this disease.

References

1. Nogawa K, Tsuritani I, Kido T, Honda R, Yamada Y, Ishizaki M. Mechanism for bone disease found in inhabitants environmentally exposed to cadmium: decreased serum 1 alpha, 25-dihydroxyvitamin D level. Int Arch Occup Environ Health. 1987;59:21–30.
2. Aoshima K, Iwata K, Kasuya M. Environmental exposure to cadmium and effects on human health. Part 2. Bone and mineral metabolism in inhabitants of the cadmium-polluted Jinzu River basin in Toyama Prefecture. Nihon Eiseigaku Zasshi. 1988;43:864–71.
3. Kazantzis G. Cadmium, osteoporosis and calcium metabolism. Biometals. 2004;17:493–8.
4. Baba H, Tsuneyama K, Kumada T, Aoshima K, Imura J. Histopathological analysis for osteomalacia and tubulopathy in itai-itai disease. J Toxicol Sci. 2014;39:91–6. (Article in Japanese)
5. Noda M, Yasuda M, Kitagawa M. Iron as a possible aggravating factor for osteopathy in itai-itai disease, a disease associated with chronic cadmium intoxication. J Bone Miner Res. 1991;6:245–55.
6. Takebayashi S, Harada T, Kamura S, Satoh T, Segawa M, Yajima K. Cadmium-induced osteopathy: clinical and autopsy findings of four patients. Appl Pathol. 1987;5:190–7.
7. Maatman RG, Van Kuppevelt TH, Veerkamp JH. Two types of fatty acid-binding protein in human kidney. Isolation, characterization and localization. Biochem J. 1991;273:759–66.
8. Prozialeck WC, Edwards JR. Mechanisms of cadmium-induced proximal tubule injury: new insights with implications for biomonitoring and therapeutic interventions. J Pharmacol Exp Ther. 2012;343:2–12.
9. Hall JE. Renal tubular reabsorption and secretion. In: Textbook of medical physiology. 13th ed; 2016. p. 353-4.

Chapter 4
Long-Term Follow-Up Study of Residents Exposed to Cadmium in Kakehashi River Basin, Ishikawa, Japan

Kazuhiro Nogawa, Yasushi Suwazono, and Teruhiko Kido

Abstract The Kakehashi River basin in Ishikawa Prefecture, Japan, was polluted with cadmium (Cd) owing to mining activities in the early twentieth century. After mining operations ceased, all polluted soil was removed and replaced by 1988. An epidemiological study in 1981–1982 showed 19.7% of Cd-exposed residents had renal tubular dysfunction by β_2-microglobulin (β_2-MG)-uria. A 28-year follow-up study found renal dysfunction was irreversible. Bone damage was also found in Cd-exposed residents with renal tubular dysfunction. Biological half-life of urinary (U)-Cd was estimated as 12.4 years in men and 11.4 years in women for unadjusted U-Cd and 16.0 years in men and 20.4 years in women for creatinine-adjusted U-Cd. Age-adjusted benchmark dose limit (BMDL) of U-Cd for β_2-MG-uria was 3.5 μg/g creatinine (Cr) in men and 3.7 μg/g Cr in women. The lowest BMDL of rice-Cd for β_2-MG-uria was 0.25 mg/kg in men and 0.24 mg/kg in women. Lifetime Cd intake was 1.3 g.

Keywords Urinary cadmium · β_2-microglobulin · Biological half-life · Long-term follow-up study · Risk assessment · Benchmark dose limit · Human

4.1 Outline of Cadmium Pollution in the Kakehashi River Basin

The Kakehashi River is about 42 km long and flows through Komatsu in the southern part of Ishikawa Prefecture. Upriver, there are mines which began operations at the end of the seventeenth century; in the river basin, there are numerous rice

K. Nogawa · Y. Suwazono
Department of Occupational and Environmental Medicine, Graduate School of Medicine, Chiba University, Chiba, Japan

T. Kido (✉)
Faculty of Health Sciences, Institute of Medical, Pharmaceutical and Health Sciences, Kanazawa University, Kanazawa, Ishikawa, Japan
e-mail: kido@mhs.mp.kanazawa-u.ac.jp

S. Himeno, K. Aoshima (eds.), *Cadmium Toxicity*, Current Topics in Environmental Health and Preventive Medicine,
https://doi.org/10.1007/978-981-13-3630-0_4

paddies. The mines were modernized at the beginning of the twentieth century and began full-scale operations in about 1930. All the mines were closed down in 1971 owing to insufficient production. When in operation, the total output of one mine was estimated to be more than 4,070,000 tons (1930). The total output of crude copper from these mines exceeded 20,000 tons per year in the most productive period (about 1960).

Paddy soil in an area of about 700 ha, extending from the middle to the lower reaches of the river, was contaminated by heavy metals from mining operations. Crops were damaged in some areas. Mining companies paid over JPY 70,000,000 (USD 233,000 in 1977) in compensation for the decreased crop yield and for the prevention of further mine pollution [1].

In 1969, 0.011 ppm cadmium (Cd) was detected in the water of the Kakehashi River. This value exceeded 0.01 ppm of the environmental water quality standard in Japan. A survey held by the Ishikawa Prefectural Government in 1974 revealed that the average concentrations of Cd and Cu were higher in the paddy soil of the areas located directly along the Kakehashi River than in the soil of other areas. Cd concentration in rice above 0.4 ppm was shown in 12 of 22 villages in the Kakehashi River basin. Starting in 1977, the polluted rice paddy soil was removed and replaced with unpolluted soil in each village. All soil removal was completed in 1988. The recent mean Cd concentration in rice decreased to 0.03 ppm in these improved areas [2].

4.2 Health Survey of the Residents

Two epidemiological studies on adverse health effect by exposure to Cd were conducted in these Cd-polluted areas [2]. The first health survey was performed in 1974–1975. Participants in the survey were 2691 residents of Cd-polluted villages and 248 residents of unpolluted villages. The participation rate was 96%. The primary examination consisted of a questionnaire and semiquantitative determination of retinol-binding protein, total protein, and glucose in early morning urine. Out of the 278 participants (10.3% of Cd-exposed participants) who showed tubular proteinuria (urinary concentration of retinol-binding protein >4 mg/L) or proteinuria with glucosuria in the primary examination, 275 underwent the secondary examination, which included the determination of total protein, glucose, creatinine, total amino nitrogen, proline, and Cd in urine. One hundred twenty-seven participants were selected for the tertiary examination, and 106 were actually examined. The examination included a medical interview, general clinical examination, orthopedic examination, X-ray of the right upper arm and shoulder, X-ray of the pubic region, urinalysis, blood chemistry, and renal function test (creatinine clearance and tubular reabsorption of phosphate (%TRP)). Finally, 39 participants (1.4%) were judged as requiring general consultation for renal tubular dysfunction, and 47 participants (1.7%) were judged as requiring clinical follow-up studies. No participants showed osteomalacic symptoms characteristic of *itai-itai* disease.

The second epidemiological study was conducted in 1981–1982 on 3178 participants (1424 men and 1754 women; 91% of the 3508 residents of the Kakehashi River basin) and 294 control participants (133 men and 161 women; 94% of the 312 residents of unpolluted villages) all of whom were over 50 years of age. It was quite different from the first epidemiological study in that β_2-MG determined by radioimmunoassay, which is considered a more sensitive indicator of renal tubular dysfunction, was included in the primary examination. Out of 627 participants (19.7% of Cd-exposed participants) who showed β_2-MG-uria (more than 1000 μg/L) or proteinuria with glucosuria or aminoaciduria, 464 participants underwent secondary examination. In total, 286 participants were selected for the tertiary examination on the basis of the following criterion from the secondary examination: %TRP of less than 80. Out of 286 participants, 234 were selected and examined. The content of the tertiary examination was the same as that of the first epidemiological study. Finally, 200 participants (6.3%) were officially recognized as having Cd-induced renal tubular dysfunction requiring observation by the research committee organized by the prefectural health authorities.

As the results of these epidemiological studies show, the prevalence of β_2-MG-uria was significantly higher in Cd-exposed participants than in unexposed participants. However, no significant difference was noted in the concurrent prevalence of proteinuria and glucosuria between the two groups [3]. Urinary metallothionein, another low-molecular-weight protein different fromβ_2-MG, was also determined in the second epidemiological study. Women living in the Cd-polluted areas were found to have a significantly higher concentration of urinary metallothionein than controls [4]. Finally, 200 participants were recognized as having renal tubular dysfunction and received annual health checkups from 1983.

4.3 Renal Effects

Urinary α_1-microglobulin was significantly higher in Cd-exposed participants than in controls [5]. Enzymes located in the brush border of renal proximal tubule are good indicators of Cd-induced renal damage different from low-molecular-weight proteins in urine. *N*-acetyl-β-D-glucosaminidase and intestinal alkaline phosphatase in urine were analyzed, and both indicators were significantly higher in Cd-exposed participants than in controls [6, 7].

Lower creatinine clearance was detected at the early stage of renal dysfunction, and a significant correlation between %TRP and glomerular filtration rate was observed. These results provide evidence of glomerular effects in people exposed to Cd. With progression of renal tubular damage, serum concentration of creatinine increased, and arterial blood pH decreased [8].

In one of the most severe cases in these follow-up studies, the patient had a serum creatinine value of 4.4 mg/100 mL, with subsequent renal failure characterized by severe anemia and metabolic acidosis.

4.4 Follow-Up Studies

The reversibility of renal tubular dysfunction was examined from 1981 to 1986, using urinary β_2-MG in 74 residents in Kanahira, the most severely polluted village in the Kakehashi River basin [9]. In cases where the level of β_2-MG in urine was 1000 μg/g Cr or greater, almost all individuals showed worsening of β_2-MG-uria. In contrast, in cases where the excretion of β_2-MG into urine was less than 1000 μg/g Cr, no significant changes were observed.

Further health examinations were carried out in 1991, 1999, 2003, 2008, and 2014. Glucose, protein, amino acid, β_2-MG, and *N*-acetyl-β-D-glucosaminidase in urine were selected as indicators of renal dysfunction. The amount of Cd in the urine was also measured as an index of Cd exposure. In 2003, the urinary (U)-Cd and β_2-MG concentrations of 25 men and 28 women did not show a significant change [10]. As results in 2008, 68 residents (32 men and 36 women) who underwent all five screenings during the 22-year period were extracted as participants. U-Cd concentrations were significantly decreased, while urinaryβ_2-MG concentrations were significantly increased. Once renal tubular dysfunction occurred after Cd exposure, irreversible aggravation was inevitable [11]. In 2014, a 28-year follow-up study was conducted to clarify the relationship between the concentration of Cd and β_2-MG in the urine of 28 residents (14 men and 14 women). All participants were over 60 years old in 2014 and participated in all six health examinations conducted over 28 years (1986–2014).U-Cd concentrations were found to have decreased by nearly half between 1986 and 2008 in both men and women, whereas it increased significantly from 2008 to 2014 in men. In contrast, urinary β_2-MG concentrations tended to increase over the 28-year study period in both sexes, as shown in Table 4.1. U-Cd and β_2-MG concentrations in women were significantly higher than those in men in this Cd-polluted region [12]. The cause of increase in U-Cd is presently examined including analysis of Cd in rice.

4.5 Bone Effects

Bone density was measured in 206 participants (92 men and 114 women) with Cd-induced renal tubular dysfunction and 110 non-exposed controls (44 men and 66 women) using microdensitometry. Bone density in Cd-exposed participants was significantly lower than that in non-exposed controls [13]. The relationship between bone density and renal function was also studied using multivariate analysis. Urinary β_2-MG was significantly associated with bone indices only in Cd-exposed participants [14]. The degree of bone damage is closely associated with the degree of renal damage. These results verified that osteopenia, which is one of the main manifestations of osteoporosis, can be detected in the early stage of chronic Cd intoxication.

Bone Gla-protein and alkaline phosphatase, which are osteoblastic indicators, were higher in Cd-exposed populations than in residents of non-polluted areas

Table 4.1 Changes in urinary cadmium (Cd) and β2-microglobulin(β2-MG) concentrations (Cr-adjusted and uncorrected) from 1986 to 2014 in males and females

	U-Cd/Cr (μg/g.cr)						U-Cd (μg/l)					
	Males = 14			Females = 14			Males = 14			Females		
Year	GM	GSD	p	GM	GSD	p	GM	GSD	p	GM	GSD	p
1986	5.45	1.38	0.02	7.46	1.61		5.95	1.69	0.0015	4.62	2.44	0.005
1991	4.18	1.51	0.04	7.29	1.64	0.016	3.70	2.14	0.0026, 0.006	5.83	2.05	0.027, 0.023
1999	2.75	1.63	< 0.0001, 0.004	4.77	1.90	0.022	2.90	2.13	< 0.001, 0.0007	3.42	2.03	0.0006, 0.008
2003	2.89	1.94		4.62	1.68		2.35	2.59	0.009	2.82	1.94	0.02
2008	1.90	2.3		3.68	1.72		1.46	2.04		1.59	2.26	
2014	3.65	1.61	0.032	4.91	1.35		2.96	2.26	0.023	2.46	2.27	

	U-β2-MG/Cr (μg/g.cr)						U-β2-MG (μg/l)					
	Males = 14			Females = 14			Males = 14			Females = 14		
Year	GM	GSD	p	GM	GSD	p	GM	GSD	p	GM	GSD	p
1986	67.3	5.64		154.7	5.63		73.4	5.58		95.8	4.46	
1991	78.8	2.26		227.5	4.23		69.7	2.12		182.1	3.64	
1999	13.6	12.9	0.018	163.2	13.7		14.3	9.52	0.03	117.2	9.04	
2003	122.6	4.48	0.002, 0.0002	296.3	4.70		99.8	2.97	0.004	181.2	4.02	
2008	196.2	4.40		760.8	8.50		150.9	3.97	0.0002	328.7	6.63	
2014	312.7	7.90		801.2	13.3		253.8	8.92		401.8	11.3	

GM geometric mean, *GSD* geometric standard deviation, *P* *p*-value

[15, 16]. An assay for the measurement of the bone resorption marker, Type 1 collagen cross-linked N-telopeptides (NTx), was produced directly from active osteoclasts at the bone surface. The amount of NTx produced is related to bone resorption activity. Mean urinary NTx concentrations of Cd-exposed women were significantly higher than those in non-exposed women; no significant difference was shown between men in the two groups [17].

Three autopsied cases have been reported among the Cd-exposed participants with renal tubular dysfunction. All had a mild or moderate degree of osteomalacia, although Looser zones were not found on radiographs.

4.6 Biological Half-Life of Urinary Cd

U-Cd is recognized as a useful marker for the total body burden of Cd exposure [18]. In terms of the absorption and elimination of Cd in humans, one-compartment models of Cd metabolism have been adopted [19, 20]. As a result, the whole-body half-life of Cd in humans is estimated to be at least 20 years. However, the attenuation and biological half-life of U-Cd have seldom been reported in the general population.

Ishizaki et al. investigated the biological half-life of U-Cd in residents of the Kakehashi River basin based on a 22-year follow-up study after cessation of environmental Cd pollution [21]. Spot urine samples were collected from residents (32 men and 36 women) in the target area in 1986, 1991, 1999, 2003, and 2008. These residents were divided into higher or lower baseline U-Cd group by the cutoff of 5 μg/L or 5 μg/g Cr. Residents whose U-Cd exceeded these cutoff values were considered to have been obviously exposed to Cd, because in Ishikawa Prefecture, the geometric mean of U-Cd grouped according to age did not exceed these values. Biological half-life of U-Cd was estimated using a linear mixed model adjusted for the baseline age.

Table 4.2 shows the baseline characteristics and age-adjusted biological half-life of U-Cd grouped according to baseline U-Cd level and sex by the linear mixed model. In the higher baseline U-Cd groups, the estimated half-life and 95% confidence intervals were 12.4 years (9.3–18.8 years) in men and 11.4 years (9.3–14.6 years) in women for unadjusted U-Cd. For creatinine-adjusted U-Cd, estimated half-life and 95% confidence intervals were 16.0 years (13.0–20.7 years) in men and 20.4 years (16.6–26.2 years) in women. In the lower baseline U-Cd groups, biological half-life for unadjusted U-Cd in men was significant (23.4 years) and longer than the corresponding half-life in the higher baseline U-Cd group.

U-Cd half-life estimated in a previous study [22] similarly was 13.6–14.2 years in men and 13.9–23.5 years in women, demonstrating the consistency of the results in these studies. However, in this study, the 95% confidence intervals were generally narrower than those in the previous study [22]. Therefore, the accuracy of the obtained biological half-life of U-Cd was improved owing to the increased sample size and improved statistical method of the linear mixed model. In this model, each

Table 4.2 Baseline characteristics and age-adjusted biological half-life of urinary cadmium (U-Cd) grouped according to baseline U-Cd level and sex by the linear mixed model [20]

			Baseline U-Cd		Baseline age (years)		Estimated biological half-life	
		N	GM	GSD	Mean	SD	(95%CI)	*P*
U-Cd (μg/L)								
0–4.9	Men	15	2.3	2.5	39.7	19.5	23.4 (14.4, 62.4)	0.002
	Women	15	2.3	1.9	40.9	14.8	NA	0.377
5.0–	Men	17	8.0	1.6	45.5	9.0	12.4 (9.3, 18.8)	<0.001
	Women	21	8.5	1.5	49.0	10.8	11.4 (9.3, 14.6)	<0.001
U-Cd (μg/g Cr)								
0–4.9	Men	14	2.0	2.8	35.8	17.3	NA	0.054
	Women	10	2.6	1.7	38.9	16.3	NA	0.916
5.0–	Men	18	6.8	1.3	48.2	10.2	16.0 (13.0, 20.7)	<0.001
	Women	26	9.3	1.4	48.3	10.9	20.4 (16.6, 26.2)	<0.001

Note: *SD* standard deviation, *GM* geometric mean, *GSD* geometric standard deviation, *NA* not available because of insignificant results, *P* *p*-value of estimated slope for observation period by linear mixed model

participant was assumed to have an underlying level of response, which was incorporated by regarding this participant effect as random [23]. We consider that the linear mixed model by taking this random participant effect into consideration would compensate, to some extent, for the absence of detailed data on lifestyle, educational, and socioeconomic factors. Therefore in this study, a more accurate and reliable biological half-life of U-Cd was estimated in the residents of a Cd-polluted area of Japan by applying updated efficient statistical methods to a larger number of participants than our previous study, based on a long observation after cessation of Cd pollution. The biological half-lives of U-Cd obtained in this study were identical to values for U-Cd or total body burden determined by a different method.

4.7 The Benchmark Dose (BMD) Method of U-Cd, Rice Cd Concentration, and Lifetime Cd Intake for Renal Effects

Recently, the benchmark dose (BMD) method has been applied to estimate the reference point of U-Cd, rice Cd concentration (R-Cd), and lifetime Cd intake (LCd) for renal effects. The BMD is defined as the exposure level which corresponds to a certain increase in the probability of an adverse response, compared with zero-background exposure. The lower 95% confidence limit of the BMD (BMDL) can be used in risk assessment as a replacement for the no observed adverse effect level [24].One notable feature of the BMD/BMDL approach is that it reflects the whole dose–response curve [24]. Shimizu et al. estimated BMDL using data from the Kakehashi River basin [25]. The study participants consisted of residents aged

50 years or more from a non-polluted area (n = 294) or from the polluted area (n = 3178) in Kakehashi River basin. BMDL for tubular β_2-MG-uria was 2.9–4.0 μg/g Cr in men and 1.5–3.6 μg/g Cr in women. Kobayashi et al. [26] also studied residents aged 50 years or more from a non-polluted area (n = 2929) or from the polluted area in the Kakehashi River basin (n = 3103). The authors estimated age-adjusted BMDL using a multiple logistic regression models including age as a covariate. The obtained BMDL was 2.7–4.6 μg/g Cr in men and 3.2–6.3 μg/g Cr in women, assuming a mean age of 64 years. Assuming a population aged 75 years or older, BMDL was 1.6–2.4 μg/g Cr in men and 1.7–2.6 μg/g Cr in women. Furthermore, Suwazono et al. estimated age-adjusted BMDL of U-Cd using the hybrid method [27]. By this method, the BMD and BMDL were estimated based on a continuous measure of the exposure and the outcome, avoiding categorization of participants with respect to the effect or to the exposure [28, 29]. The study participants consisted of residents of a non-polluted area (n = 1509) and of the Kakehashi River basin (n = 3103) aged 50 years or older. Assuming mean age of 63 years, obtained BMDL of U-Cd for β_2-MG-uria was 3.5 μg/g Cr in men and 3.7 μg/g Cr in women [27]. These values were consistent with those in the study by Kobayashi et al. [26]. In the present study, the risk was assessed using continuous effect markers and the hybrid approach. Considering all of these studies, the reference point of U-Cd was thought to be 3.5–3.7 in the Kakehashi River basin.

Nogawa et al. estimated the BMDL of R-Cd for various renal effects by applying the hybrid approach [30]. The participants consisted of residents aged 50 years or more living in the Kakehashi River basin for at least 30 years (1120 men, 1274 women). R-Cd was determined for rice samples stored in warehouses of the farmers in all of the polluted villages. The obtained BMDLs were 0.39 (amino nitrogen), 0.26 (metallothionein), and 0.25 (β_2-MG) mg/kg in men and 0.44 (glucose), 0.32 (protein), 0.33 (amino nitrogen), 0.28 (metallothionein), and 0.24 (β_2-MG) mg/kg in women. The lowest BMDL was 0.25 (β_2-MG) mg/kg in men and 0.24 (β_2-MG) mg/kg in women. These values were lower than the maximum level (0.4 mg/kg) determined by the Codex Alimentarius Commission.

The reference level of LCd was estimated as the BMD and BMDL for various renal effects by applying a hybrid approach [31]. The participants consisted of residents of non-polluted areas (n = 278) and of the Kakehashi River basin (n = 3013). Glucose, protein, amino nitrogen, metallothionein, and β_2-MG in urine were measured as indicators of renal dysfunction. The BMD and BMDL that corresponded to an additional risk of 5% were calculated with background risk at zero exposure set at 5%. The obtained BMDLs of LCd were 3.7 (glucose), 3.2 (protein), 3.7 (amino nitrogen), 1.7 (metallothionein), and 1.8 (β_2-MG) g in men and 2.9 (glucose), 2.5 (protein), 2.0 (amino nitrogen), 1.6 (metallothionein), and 1.3 (β_2-MG) g in women. The lowest BMDL was 1.7 (metallothionein) g in men and 1.3 (β_2-MG) g in women. The lowest BMDL of LCd (1.3 g) was somewhat lower than the representative threshold LCd (2.0 g) calculated in previous studies [32–34]. Several years ago, the Panel on Contaminants in the Food Chain (CONTAM) of the European Food Safety Authority (EFSA) established a tolerable weekly intake (TWI) for Cd of 2.5 μg/kg body weight [35]. Furthermore, the

Joint FAO/WHO Expert Committee on Food Additives (JECFA) had set a provisional TWI of 7 μg/kg body weight but corrected the provisional tolerable monthly intake (PTMI) of 25 μg/kg body weight, which corresponds to a provisional TWI of 5.8 μg/kg body weight [36]. In contrast, with the maximum level of R-Cd of 0.4 mg/kg adopted by the Codex Alimentarius Commission, assuming the person has eaten 150–350 g of rice per day for 50 years, the total Cd intake from rice would be 1.10–2.56 g. These are similar to or higher than those calculated BMDL of LCd in the present study. As noted above, the biological half-life of U-Cd has been estimated to be 12–20 years [20], suggesting that a long period is required to decrease the health burdens associated with Cd in the population and highlighting that reduction of environmental Cd exposure is an important and urgent issue in the general population. The obtained BMDLs for U-Cd, R-Cd, and LCd may contribute to further discussion on the health risk assessment of Cd exposure.

4.8 Conclusion

The prevalence of Cd-induced renal tubular dysfunction was about 20%, using urinary β_2-MG of residents of the Kakehashi River basin over 50 years of age. Renal dysfunction was irreversible and bone damage was also found in Cd-exposed participants with renal tubular dysfunction. Using the linear mixed model, a more reliable biological half-life of U-Cd was estimated as 12.4 years in men and 11.4 years in women for unadjusted U-Cd and 16.0 years in men and 20.4 years in women for creatinine-adjusted U-Cd. Using the hybrid approach, age-adjusted BMDL of U-Cd for β_2-MG-uria was 3.5 μg/g Cr in men and 3.7 μg/g Cr in women. The lowest BMDL of R-Cd for β_2-MG-uria was 0.25 mg/kg in men and 0.24 mg/kg in women. The lowest BMDL of LCd (1.3 g) was somewhat lower than the representative threshold LCd (2.0 g) calculated in previous studies.

References

1. Kawano S, Kato T, Ishikawa Prefecture, epidemiological studies. In: Tsuchiya K, editor. Cadmium studies in Japan: a review. Tokyo: Kodansha; 1978. p. 220–5.
2. Kido T. Environmental cadmium pollution and its health effects on inhabitants in Japan The Kakehashi River Basin, Ishikawa Prefecture. In: Nogawa K, Kurachi M, Kasuya M, editors. Advances in the prevention of environmental cadmium pollution and countermeasures. Kanazawa: Eiko Laboratory; 1999. p. 37–42.
3. Nordberg G, Kido T, Roels HA. Cadmium-induced renal effects. In: ME DB, Porter GA, editors. Clinical nephrotoxins renal injury from drugs and chemicals. 3rd ed. New York: Springer; 2008. p. 785–810.
4. Shaikh ZA, Kido T, Kito H, Honda R, Nogawa K. Prevalence of metallothioneinuria among the population living in the Kakehashi River basin in Japan - an epidemiological study. Toxicology. 1990;64:59–69.

5. Nogawa K, Kido T, Yamada Y, Tsuritani I, Honda R, Ishizaki M, Terahata K. α_1-microglobulin in urine as an indicator of renal tubular damage caused by environmental cadmium exposure. Toxicol Lett. 1984;22:63–8.
6. Kido T. Renal and bone effects. Environ Sci. 1995;3:149–62.
7. Kido T, Kobayashi E, Hayano M, Nogawa K, Tsuritani I, Nishijo M, Tabata M, Nakagawa H, Nuyts GD, DeBroe ME. Significance of elevated urinary human intestinal alkaline phosphatase in Japanese people exposed to environmental cadmium. Toxicology. 1995;80:49–54.
8. KidoT NK, Ishizaki M, Honda R, Tsuritani I, Yamada Y, Nakagawa H, Nishi M. Long-term observation of serum creatinine and arterial blood pH in persons with cadmium-induced renal dysfunction. Arch Environ Health. 1990;45:35–41.
9. Kido T, Honda R, Tsuritani I, Yamaya H, Ishizaki M, Yamada Y, Nogawa K. Progress of renal dysfunction in inhabitants environmentally exposed to cadmium. Arch Environ Health. 1988;43:213–7.
10. Sato R, Kido T, Honda R, Nishijo M, Nakagawa H, Kobayashi E, Suwazono Y. Seventeen-year observation on urinary cadmium and β_2-microglobulin in inhabitants after cessation of cadmium-exposure in Japan. Bull Environ Contam Toxicol. 2010;84(4):363–7.
11. Sato R, Kido T, Nakagawa H, Nishijo M, Honda R, Kobayashi E, Suwazono Y. Twenty-two-year observation on urinary cadmium and ß$_2$-microglobulin in inhabitants after cessation of cadmium-exposure in Japan. OJNeph. 2013;3(4):205–10.
12. Phuc HD, Kido T, Manh HD, Anh LT, Oanh NTP, Okamoto R, Ichimori A, Nogawa K, Suwazono Y, Nakagawa H. A 28-year observational study of urinary cadmium and β_2-microglobulin concentrations in inhabitants in cadmium-polluted areas in Japan. J Appl Toxicol. 2016;36:1622–8.
13. Kido T, Nogawa K, Yamada Y, Honda R, Tsuritani I, Ishizaki M, Yamaya H. Osteopenia in inhabitants with renal dysfunction induced by exposure to environmental cadmium. Int Arch Occup Environ Health. 1989;61:271–6.
14. Kido T, Nogawa K, Honda R, Tsuritani I, Ishizaki M, Yamada Y, Nakagawa H. The association between renal dysfunction and osteopenia in environmental cadmium-exposed subjects. Environ Res. 1990;51:71–82.
15. Kido T, Honda R, Tsuritani I, Ishizaki M, Yamada Y, Nakagawa H, Nogawa K, Dohi Y. Serum levels of bone Gla-Protein in inhabitants exposed to environmental cadmium. Arch Environ Health. 1991;46(1):43–9.
16. Tsuritani I, Honda R, Ishizaki M, Yamada Y, Aoshima K, Kasuya M. Serum bone-type alkaline phosphatase activity in woman living in a cadmium-polluted area. Toxicol Lett. 1994;71:209–16.
17. Omote S, Kido T, Nishijo M, Nakagawa H, Suwazono Y, Kobayashi E, Nogawa N. Urinary type I collagen cross-linked N-telopeptides in inhabitants 18 years after cessation of exposure to cadmium in Japan. Bull Environ Contam Toxicol. 2006;76(2):184–94.
18. Järup L, Åkesson A. Current status of cadmium as an environmental health problem. Toxicol Appl Pharmacol. 2009;238(3):201–8.
19. Kjellström T. A mathematical model for the accumulation of cadmium in human kidney cortex. Nord Hyg Tidskr. 1971;52(2):111–9.
20. Tsuchiya K, Sugita M. A mathematical model for deriving the biological half-life of a chemical. Nord Hyg Tidskr. 1971;52(2):105–10.
21. Ishizaki M, Suwazono Y, Kido T, Nishijo M, Honda R, Kobayashi E, et al. Estimation of biological half-life of urinary cadmium in inhabitants after cessation of environmental cadmium pollution using a mixed linear model. Food Addit Contam Part A Chem Anal Control Expo Risk Assess. 2015;32(8):1273–6.
22. Suwazono Y, Kido T, Nakagawa H, Nishijo M, Honda R, Kobayashi E, et al. Biological half-life of cadmium in the urine of inhabitants after cessation of cadmium exposure. Biomarkers. 2009;14(2):77–81.
23. Fitzmaurice GM, Laird NM, Ware JH. Applied longitudinal analysis. Hoboken, NJ: John Wiley & Sons; 2004.

24. EPA US. The use of the benchmark dose (BMD) approach in health risk assessment. Final report. Washington, DC: Risk Assessment Forum, U.S. Environmental Protection Agency; 1995.
25. Shimizu A, Kobayashi E, Suwazono Y, Uetani M, Oishi M, Inaba T, et al. Estimation of benchmark doses for urinary cadmium based on beta2-microglobulin excretion in cadmium-polluted regions of the Kakehashi River basin, Japan. Int J Environ Health Res. 2006;16(5):329–37.
26. Kobayashi E, Suwazono Y, Dochi M, Honda R, Nishijo M, Kido T, et al. Estimation of benchmark doses as threshold levels of urinary cadmium, based on excretion of beta2-microglobulin in cadmium-polluted and non-polluted regions in Japan. Toxicol Lett. 2008;179(2):108–12.
27. Suwazono Y, Nogawa K, Uetani M, Nakada S, Kido T, Nakagawa H. Application of the hybrid approach to the benchmark dose of urinary cadmium as the reference level for renal effects in cadmium polluted and non-polluted areas in Japan. Environ Res. 2011;111(2):312–4.
28. Crump K. Calculation of benchmark doses from continuous data. Risk Anal. 1995;15(1):79–89.
29. Sand S, Victorin K, Filipsson AF. The current state of knowledge on the use of the benchmark dose concept in risk assessment. J Appl Toxicol. 2008;28(4):405–21.
30. Nogawa K, Kido T, Nishijo M, Nakagawa H, Suwazono Y. Benchmark dose of cadmium concentration in rice for renal effects in a cadmium-polluted area in Japan. J Appl Toxicol. 2015;35(1):24–8.
31. Kubo K, Nogawa K, Kido T, Nishijo M, Nakagawa H, Suwazono Y. Estimation of benchmark dose of lifetime cadmium intake for adverse renal effects using hybrid approach in inhabitants of an environmentally exposed river Basin in Japan. Risk Anal. 2017;37(1):20–6.
32. Kobayashi E, Suwazono Y, Dochi M, Honda R, Kido T. Association of lifetime cadmium intake or drinking Jinzu River water with the occurrence of renal tubular dysfunction. Environ Toxicol. 2009;24(5):421–8.
33. Chiyoda N, Kobayashi E, Okubo Y, Suwazono Y, Kido T, Nogawa K. Allowable level of lifetime cadmium intake calculated from the individuals in the Jinzu River basin of Japan. Biol Trace Elem Res. 2003;96(1-3):9–20.
34. Watanabe K, Kobayashi E, Suwazono Y, Okubo Y, Kido T, Nogawa K. Tolerable lifetime cadmium intake calculated from the inhabitants living in the Jinzu River basin, Japan. Bull Environ Contam Toxicol. 2004;72(6):1091–7.
35. EFSA. Scientific opinion of the panel on contaminants in the food chain on a request from the European Commission on cadmium in food. EFSA J. 2009;980:1–139.
36. FAO/WHO. The summary report of the 73rd JECFA meeting. Geneva: World Health Organization; 2010.

Chapter 5
Effects of Cadmium Exposure on Life Prognosis

Muneko Nishijo and Hideaki Nakagawa

Abstract The increased cancer incidence and mortality associated with cadmium (Cd) exposure have been investigated worldwide, but the findings are often inconsistent. The increased mortality from cardiovascular and other non-cancer diseases has also been reported in populations exposed to Cd. In Japanese Cd-polluted areas, including one where *itai-itai* disease is endemic, levels of urinary Cd and renal tubular dysfunction induced by Cd exposure are associated with an increased mortality in both sexes (women > men), particularly from cardiovascular and renal diseases. However, deaths from respiratory and digestive diseases have also been recorded in patients with *itai-itai* disease. Significant association between Cd in rice and mortality was found only in the most polluted part of the Jinzu River basin, Toyama in Japan. These findings indicate that high-level Cd exposure, whether environmental or occupational, may adversely affect life prognosis, but more studies are required to characterize the effects of low-level Cd exposure indicated by urinary or blood Cd in general populations with different dietary and lifestyle habits.

Keywords Cadmium · Mortality · Causes of deaths · Environmental exposure *Itai-itai* disease

5.1 Introduction

Associations between cadmium (Cd) exposure and cancer have been investigated in industrial settings, and increased mortality from lung cancer and prostate cancer in male workers was reported [1–5]. Based on findings from human and animal studies, the International Agency for Research on Cancer (IARC) [6, 7] concluded that occupational Cd exposure increases cancer risk.

M. Nishijo (✉)
Department of Public Health, Kanazawa Medical University, Uchinada, Ishikawa, Japan
e-mail: ni-koei@kanazawa-med.ac.jp

H. Nakagawa
Medical Research Institute, Kanazawa Medical University, Uchinada, Ishikawa, Japan

S. Himeno, K. Aoshima (eds.), *Cadmium Toxicity*, Current Topics in Environmental Health and Preventive Medicine,
https://doi.org/10.1007/978-981-13-3630-0_5

Lauwerys and De Wals [8] reported an increased mortality from renal diseases in individuals environmentally exposed to Cd in Belgium, while Shigematsu et al. [9] reported lower mortality from all causes in residents of heavily polluted and *itai-itai* disease-endemic areas in Japan. Kawano et al. [10] compared mortality from all causes in patients with *itai-itai* disease and in age- and sex-matched controls and reported significantly higher mortality in the former. In these earlier studies, standardized mortality ratios (SMRs) and Kaplan-Meier curves were commonly used, but mortality risk analysis using the Cox proportional hazards model was popularized in the 1990s to adjust for confounding factors such as age. Several studies evaluated the association between Cd exposure and increased mortality or renal tubular dysfunction in four Cd-polluted areas in Japan [11–16] and found a higher mortality in individuals with renal tubular dysfunction induced by Cd exposure. More studies were performed to identify the specific causes of death, with particular attention paid to cancer mortality rates in these areas [17, 18] and in a Cd-polluted area in Belgium [19].

Associations between urinary and blood Cd levels and mortality have been assessed in the general population, particularly in the USA, and the findings suggested that Cd exposure was a risk factor for cardiovascular and other diseases [20–22], as well as for cancer [23–25].

These studies in the exposed and general populations which reported significant associations between markers of Cd exposure and Cd-induced health effects and mortality were listed in Table 5.1. Their results were described in the following Sects. 5.2 and 5.3. Several meta-analyses of risk for specific cancers (prostate, breast, and lung) associated with Cd exposure have been conducted, combining the findings of multiple studies in occupational and nonoccupational cohorts [38–40]. These meta-analyses used incidence data as well as mortality data to analyze the association between Cd exposure and cancer risk. We therefore added their findings to the mortality reports in each of the following sections.

5.2 Cd Exposure and Mortality

5.2.1 Dietary Cd Intake and Mortality

Cd levels in rice and daily Cd intake have commonly been used as Cd exposure markers, but dietary habits across cultures and regions affect Cd intake. Sand and Becker [41] reported that 40–50% of Cd intake in a Swedish population was derived from potatoes and wheat flour. Arnich et al. [42] also found that dietary Cd intake was attributable to bread and potato products in France. However, in Asian countries, rice, a staple food, is the major source of dietary Cd intake. High Cd intake from rice (approximately 30–70% of total intake) has been reported in the general population in China, Korea, and Japan [43–45]. In a dietary survey conducted in the 1970s, when Cd pollution was endemic in Japan, the average daily Cd intake was reported as 195 μg (82.5% from rice) for residents of Cd-polluted areas [46]. A high

Table 5.1 Studies showed significant associations between markers of Cd exposure and Cd-induced health effects and mortality

Cd exposure and Cd-induced health effect		Population	Outcome	References
Cd intake	Cd in rice	Environmentally exposed	Increased mortality for all causes	[26, 27]
Internal exposure marker	Urinary Cd	Environmentally exposed	Increased mortality for all causes, cardiovascular diseases, and cancer	[18, 19, 28, 29]
		General	Increased mortality for all causes, cardiovascular diseases, liver diseases, and cancer	[20–24], [59, 60]
	Blood Cd	Environmentally exposed	Increased mortality for cancer and cardiovascular diseases	[30]
		General	Increased mortality for all causes, cardiovascular diseases, and Alzheimer's disease	[22, 31, 32]
Cd-induced health effects	Renal tubular dysfunction markers	Environmentally exposed	Increased mortality for all causes, cardiovascular diseases, and renal disease	[11–16, 33–35]
	Itai-itai disease	Environmentally exposed	Increased mortality for all causes, pneumonia, gastrointestinal diseases, cardiovascular diseases, and renal disease	[10, 36, 37]
		General	Increased mortality for all causes	[65]

Cd contribution from rice of up to 70% was still recorded in some polluted areas in the 1990s [47]. Significant associations between Cd in rice and health effect indices such as *itai-itai* disease and renal tubular dysfunctions have been shown in Cd-polluted areas in Japan [48–50], indicating that rice is a critical element of Cd exposure in Japan.

Ishihara et al. [26] surveyed the Jinzu River basin in Japan, where *itai-itai* disease is endemic, and reported that the mortality risk of men living in areas with Cd-contaminated rice (Cd $\geq$ 0.3 ppm) was significantly higher than for inhabitants of areas where rice contained <0.3 ppm Cd. Kobayashi et al. [27] reported an increased mortality risk in women from the same area, which was associated with total Cd intake from rice. However, no significant associations between Cd-polluted rice and mortality have been reported in other Cd-polluted areas in Japan, suggesting that only in the most polluted part of the Jinzu River basin would Cd in rice affect the mortality rate.

Several studies have been performed in general populations to evaluate the association between daily Cd intake and the incidence of hormone-related cancers, but to our knowledge, no mortality study has been published. Dietary Cd intake increased the incidence of endometrial cancer [51] and breast cancer [52] in postmenopausal women in Sweden. Among postmenopausal Japanese women, a significant association between dietary Cd intake and estrogen receptor-positive breast cancer was found in a cohort of 405 matched case-control pairs at 4 hospitals in Nagano Prefecture [53]. However, a meta-analysis of observational studies on dietary cadmium intake and breast cancer did not find a significant association [40]. In a cohort of Danish men, no association was found between dietary Cd intake and prostate cancer risk [54].

5.2.2 Urinary Cd Levels and Mortality

Urinary Cd concentrations are used as an index of Cd exposure, because they reflect Cd body burden and are proportional to the renal Cd concentrations [55]. Therefore, urinary Cd levels are often used in health risk assessments of chronic Cd exposure. Associations between urinary Cd levels and mortality have been reported in residents of Cd-polluted areas in Japan [18, 28, 29] and Belgium [8, 19]. In studies of the general population, conducted mainly in the USA, the association between Cd exposure and mortality risk was assessed using urinary Cd levels [20–24].

In 15- and 22-year follow-up surveys of residents of the Cd-polluted Kakehashi River basin, Japan, an increased risk ratio for mortality from all causes associated with urinary Cd levels was reported that was concentration-dependent in both sexes (women > men). Cardiovascular conditions, primarily heart failure, and cerebral infarctions contributed to the increased mortality [18, 28, 29]. Inskip et al. [56] reported the findings of a 40-year follow-up study conducted in a UK village with Cd-contaminated soil; the study found no evidence of increased SMRs compared with a general UK population. Elliott et al. [57] reported that a higher SMR of borderline significance was evident for cardiovascular and renal diseases in the same Cd-exposed cohort, but mortality from all causes was lower than expected. Lauwerys and De Wals [8] reported high proportional mortality rates from renal diseases in a Cd-exposed Belgian cohort compared with a general population but no increase in all-cause SMRs. Nawrot et al. [19] reported a higher incidence of lung cancer (95% of cases were fatal) associated with increased urinary Cd levels in a Belgian cohort followed between 1985 and 2004 and suggested that aspiration of house dust containing Cd-contaminated soil particles may be related to the increase in lung cancer incidence.

Menke et al. [21], mining data from the US National Health and Nutrition Examination Survey (NHANES) and following the cohort over time, found that urinary Cd levels were associated with an increased risk of death from cardiovascular causes, particularly coronary vascular diseases, in men. Similarly, Tellez-Plaza et al. [22] reported an increased mortality risk from cardiovascular causes in subjects from the NHANES cohort who had urinary Cd levels ≥0.57 μg/g creatinine. In a case-control study in a US population, McElroy et al. [25] reported that breast cancer patients had higher urinary Cd levels than matched controls. In another analysis of NHANES data, increased mortality risks for lung cancer, pancreatic cancer, and non-Hodgkin lymphoma in men and lung cancer, leukemia, ovarian, and uterine cancer in women were reported in groups with higher urinary Cd levels [23]; Lin et al. [24] calculated a higher risk for all cancers in both sexes and for lung cancer in men from NHANES data on older adults with high cadmium exposure and low zinc intake. Hyder et al. [20] reported that increased urinary Cd levels were associated with the increased mortality risk from non-cancer liver diseases, hepatic necro-inflammation, and nonalcoholic fatty liver disease in men and with hepatic necro-inflammation in women. However, in a case-cohort study using the population-based Danish Diet, Cancer and Health Cohort, no association was found between urinary Cd levels and risk of developing postmenopausal breast cancer [58].

In cohort studies conducted in a general population from non-Cd-polluted areas in Japan [59, 60], the fourth quartile of urinary Cd (≥2.9–4.6 μg/g creatinine) showed significantly increased mortality risks from all causes compared with the first quartile (<1.5 μg/g creatinine) in both men and women. These findings suggest that even in non-polluted areas in Japan, at least one quarter of residents may have a higher Cd exposure level, affecting life prognosis in both sexes.

5.2.3 Blood Cd Levels and Mortality

Cd levels in blood can reflect a recent Cd exposure; the metal is transported from the liver to other organs, including the kidney, and is often seen in factory workers who are exposed to Cd at very high concentrations [61].Decreased bone mineral density was associated with blood Cd levels in Swedish male factory workers [62]; the authors suggested that, at least in men, blood Cd levels were more closely associated with bone mineral density than urinary Cd levels.

In mortality studies, urinary Cd is typically used as an exposure marker rather than blood Cd, because it is a better reflection of the Cd body burden. However, Wang et al. [30] used blood Cd levels in Chinese inhabitants of Cd-polluted areas to correlate exposure level with mortality from cardiovascular diseases and cancer.

Tellez-Plaza et al. [22] mined the NHANES database to calculate the hazard ratios for blood Cd levels in a US general population; blood Cd levels were associated with mortality from cardiovascular causes such as coronary heart disease. Min and Min [31] looked at NHANES participants aged ≥60 years and reported that participants with blood Cd levels >0.6 μg/L were more likely to die from Alzheimer's disease than participants with lower blood Cd levels. However, the sample size was small and the association was slight. Peng et al. [63] also analyzed the association between Cd exposure and death from Alzheimer's disease in NHANES participants but used urinary Cd levels as the exposure marker; the necessity to adjust the findings for urinary creatinine prevented them from reaching a general conclusion regarding the association.

Moberg et al. [32] analyzed the data from 2920 middle-aged Swedish women and found that overall mortality risk was significantly higher in women with high blood Cd levels (> 0.51 μg/L) than in women with lower blood Cd levels (< 0.18 μg/L), although specific causes of death were not reported.

5.3 Cd-Induced Renal Tubular Dysfunction and Mortality

5.3.1 Renal Tubular Dysfunction Markers and Mortality in Japan

In Cd-polluted areas in Ishikawa, Nagasaki, and Akita prefectures, several follow-up studies have been conducted to assess the association of renal tubular dysfunction and mortality. Tubular dysfunction was indicated using biomarkers such as urinary beta-2 microglobulin (β2-MG; [11, 14]), retinol-binding protein [16], amino acids [12], and a combination of marker proteins, glucose, β2-MG, and amino acids [15]. These studies reported significantly increased hazard ratios in residents of both sexes with renal tubular dysfunction. Nakagawa et al. [14] showed that the hazard ratios increased as urinary β2-MG levels increased, suggesting the usefulness of urinary β2-MG as a prognostic factor, although urinary protein has also been shown as a good marker in women [15].

Mortality analyses for specific causes of death found increased hazard ratios for renal diseases and heart failure among subjects of both sexes with renal tubular dysfunction in the Kakehashi River basin, Ishikawa Prefecture [16, 33]. Arisawa et al. [34] reported that the mortality rate from non-cancer causes among residents of a Cd-polluted area in Nagasaki Prefecture whose urinary β2-MG ≥ 1000 μg/g creatinine was significantly higher than that of the general Japanese population. However, within the Cd-polluted area, the age-adjusted mortality ratio for cancer deaths was significantly associated with increased urinary β2-MG levels.

In the Jinzu River basin, Toyama Prefecture, an area where *itai-itai* disease is endemic, a long follow-up survey found that the hazard ratios of subjects with proteinuria and/or glucosuria increased significantly in both sexes [13]. To clarify the causes of death in these subjects, Maruzeni et al. [35] performed a 26-year follow-up study of residents who participated in a health impact survey conducted by the Ministry of Environment between 1976 and 1984 and reported that among the 267 women suspected of having renal tubular dysfunction based on the presence of glucoproteinuria (proteinuria and glucosuria), mortality from all causes increased significantly. Respiratory, renal, and cardiovascular diseases, as well as other causes, with the exception of cerebrovascular diseases, contributed to the increased mortality of subjects with renal tubular dysfunction. Although renal effects induced by Cd were indicated by different markers, the mortality risks of the most affected groups in Toyama [35] and Ishikawa [14] were both increased and similar (hazard ratios of 2.8 and 2.7, respectively). Increased hazard ratios were also observed in other groups with fewer renal effects (hazard ratios of 2.4 and 2.3, respectively, for the second affected groups, and hazard ratios of 1.5 and 1.9, respectively, for the third affected groups). These findings indicate that Cd-induced renal tubular dysfunction can increase mortality risk in a concentration-dependent manner.

Suwazono et al. [64] reported increased hazard ratios corresponding to an increase in renal tubular dysfunction markers in Japanese men and women living in areas not polluted by Cd. This suggests that low-level renal tubular dysfunction resulting from Cd exposure may also affect mortality.

5.3.2 Itai-Itai *Disease and Mortality*

Itai-itai disease is caused by a Cd exposure, mainly in women, and is characterized by osteomalacia with severe bone pain and renal tubular dysfunction. This disease was first recognized in the Jinzu River basin in Toyama in the 1960s [65–67]. As of December 28, 1999, 410 Japanese patients were officially diagnosed with or suspected of having *itai-itai* disease [68]. Between 2000 and 2010, 20 new patients were officially recognized [69], although their disease was suspected to have been caused by an increased Cd body burden from exposure prior to 2000.

Kawano et al. [10] surveyed patients with *itai-itai* disease 15 years after they were diagnosed in 1967–1968 and reported a decreased survival rate compared with residents of the same area with neither proteinuria nor glucosuria. Nakagawa et al. [36] also reported an increased mortality and a shortened life span in patients with *itai-itai* disease and residents suspected of having the disease recognized in 1967–1968, in a 20-year follow-up survey.

Nishijo et al. [37] performed a nested case-control analysis of the Jinzu River basin study conducted by Maruzeni et al. [35], to evaluate the causes of death among female patients with *itai-itai* disease; the primary cohort had included only five male patients. The hazard ratios for specific causes of death of 142 female patients with *itai-itai* disease were compared with age- and occupation-matched female controls. Vital statistics data were used to determine the cause of death, but direct causes of death from autopsy records were corrected in 29 patients who had died from Cd poisoning. Among patients with *itai-itai* disease, respiratory diseases and digestive diseases contributed to the increased mortality. Renal disease contributed to mortality in these patients but also exerted a notable impact on the mortality of women with renal tubular dysfunction who had not been diagnosed with *itai-itai* disease. Although risk of death from colorectal cancer was higher in residents with renal tubular dysfunction, the increased mortality risk for all malignant neoplasms, including colorectal cancer, was not significant in patients with *itai-itai* disease suffering from the most severe chronic Cd poisoning [37].

5.4 Summary and Conclusion

Many studies on Cd toxicity and mortality have been performed in industrial workers, residents of Cd-polluted areas, and general cohorts in Europe, the USA, China, and Japan. In occupational Cd exposure, mortality from cancer, particularly lung and prostate cancer, may be increased in men. Environmental Cd exposure has yielded significant evidence of poor life prognosis in individuals of both sexes with renal tubular dysfunction, although the observed impact was stronger in women than in men. Cardiovascular diseases and renal diseases may contribute to the increased mortality, but the contribution of Cd to cancer mortality may be limited. Specific causes of death in patients with *itai-itai* disease suffering from the most severe chronic Cd poisoning were pneumonia and gastrointestinal conditions related to the illness.

In general populations and in residents of Cd-polluted areas who do not show renal effects, increased cancer incidence and mortality, particularly hormone-related cancers such as breast and prostate cancer, is a matter of concern, but the findings are inconsistent across studies. Significant association between Cd in rice and mortality was found only in the most polluted part of the Jinzu River basin, Toyama in Japan, suggesting that dietary Cd intake might be not powerful enough to show associations with incidence and mortality. Investigations into cancer incidence would be required to clarify the contribution of low-level Cd exposure indicated by urinary or blood Cd to the increased cancer risk.

References

1. Lemen RA, Lee JS, Wagoner JK, et al. Cancer mortality among cadmium production workers. Ann N Y Acad Sci. 1976;271:273–9.
2. Armstrong BG, Kazantzis G. Prostatic cancer and chronic respiratory and renal disease in British cadmium workers: a case control study. Br J Ind Med. 1985;42:540–5.
3. Elinder CG, Kjellstrom T, Hogstedt C, et al. Cancer mortality of cadmium workers. Br J Ind Med. 1985;42:651–5.
4. Ades AE, Kazantzis G. Lung cancer in a non-ferrous smelter: the role of cadmium. Br J Ind Med. 1988;45:435–42.
5. Sorahan T, Esmen NA. Lung cancer mortality in UK nickel-cadmium battery workers, 1947-2000. Occup Environ Med. 2004;61:108–16.
6. International Agency for Research on Cancer. Cadmium and cadmium compounds. In: IARC monograph, vol. 58. Lyon: International Agency for Research on Cancer; 1993. p. 119–237.
7. International Agency for Research on Cancer. Cadmium and cadmium compounds. In: IARC monograph, vol. 100C. Lyon: International Agency for Research on Cancer; 2012. p. 121–45.
8. Lauwers R, Wals D. Environmental pollution by cadmium and mortality from renal diseases. Lancet. 1981;1:383.
9. Shigematsu I, Minowa M, Usui T, et al. Epidemiological survey for causes of death among residents in cadmium polluted areas. Kankyo Hoken Report. 1981;46:1–71.
10. Kawano S, Nakagawa H, Okumura Y, et al. A mortality study of patients with *itai-itai* disease. Environ Res. 1986;40:98–102.

11. Iwata K, Saito H, Nakano A. Association between cadmium-induced renal dysfunction and mortality: further evidence. Tohoku J Exp Med. 1991;164(4):319–30.
12. Iwata K, Saito H, Moriyama M, et al. Follow up study of renal tubular dysfunction and mortality in residents of an area polluted with cadmium. Br J Ind Med. 1992;49(10):736–7.
13. Matsuda T, Kobayashi E, Okubo Y, et al. Association between renal dysfunction and mortality among inhabitants in the region around Jinzu River basin polluted by cadmium. Environ Res. 2002;88:156–3.
14. Nakagawa H, Nishijo M, Morikawa Y, et al. Urinary β2-microglobulin concentration and mortality in a cadmium-polluted area. Arch Environ Health. 1993;48:428–35.
15. Nishijo M, Nakagawa H, Morikawa Y, et al. Prognostic factors of renal dysfunction induced by environmental cadmium pollution. Environ Res. 1994;64:112–21.
16. Nishijo M, Nakagawa H, Morikawa Y, et al. Mortality of inhabitants in an area polluted by cadmium: 15 year follow up. Occup Environ Med. 1995;52:181–4.
17. Arisawa K, Nakano A, Saito H, et al. Mortality and cancer incidence among a population previously exposed to environmental cadmium. Int Arch Occup Environ Health. 2001;74(4):255–62.
18. Nishijo M, Morikawa Y, Nakagawa H, et al. Causes of death and renal tubular dysfunction in residents exposed to cadmium in the environment. Occup Environ Med. 2006;63:545–50.
19. Nawrot T, Plusquin M, Hogervorst J, et al. Environmental exposure to cadmium and risk of cancer: a prospective population-based study. Lancet Oncol. 2006;7:119–26.
20. Hyder O, Chung M, Cosgrove D, et al. Cadmium exposure and liver disease among US adults. J Gastrointest Surg. 2013;17(7):1265–73.
21. Menke A, Muntner P, Silbergeld EK, et al. Cadmium levels in urine and mortality among U.S. adults. Environ Health Perspect. 2009;117:190–6.
22. Tellez-Plaza M, Navas-Acien A, Menke A, et al. Cadmium exposure and all-cause and cardiovascular mortality in the U.S. general population. Environ Health Perspect. 2012;120:1017–22.
23. Adams SV, Passarelli MN, Newcomb PA. Cadmium exposure and cancer mortality in the Third National Health and Nutrition Examination Survey cohort. Occup Environ Med. 2012;69(2):153–6.
24. Lin YS, Caffrey JL, Lin JW, et al. Increased risk of cancer mortality associated with cadmium exposures in older Americans with low zinc intake. J Toxicol Environ Health A. 2013;76(1):1–15.
25. McElroy JA, Shafer MM, Trenthan-Dietz A, et al. Cadmium exposure and breast cancer risk. J Natl Cancer Inst. 2006;98:869–73.
26. Ishihara T, Kobayashi E, Okubo Y, et al. Association between cadmium concentration in rice and mortality in the Jinzu River basin, Japan. Toxicology. 2001;163(1):23–8.
27. Kobayashi E, Okubo Y, Suwazono Y, et al. Association between total cadmium intake calculated from the cadmium concentration in household rice and mortality among inhabitants of the cadmium-polluted Jinzu River basin of Japan. Toxicol Lett. 2002;129(1–2):85–91.
28. Nakagawa H, Nishijo M, Morikawa Y, et al. Urinary cadmium and mortality among inhabitants of a cadmium-polluted area in Japan. Environ Res. 2006;100:323–9.
29. Li Q, Nishijo M, Nakagawa H, et al. Dose-response relationship between urinary cadmium and mortality in habitants of a cadmium-polluted area: a 22-year follow-up study in Japan. Chinese Med. 2011;J124:3504–9.
30. Wang M, Xu Y, Pan S, et al. Long-term heavy metal pollution and mortality in a Chinese population: an ecologic study. Biol Trace Elem Res. 2011;142(3):362–79.
31. Min JY, Min KB. Blood cadmium levels and Alzheimer's disease mortality risk in older US adults. Environ Health. 2016;15(1):69. https://doi.org/10.1186/s12940-016-0155-7.
32. Moberg L, Nilsson PM, Samsioe G, et al. Increased blood cadmium levels were not associated with increased fracture risk but with increased total mortality in women: the Malmö Diet and Cancer Study. Osteoporos Int. 2017;28(8):2401–8. https://doi.org/10.1007/s00198-017-4047-7.
33. Nakagawa H, Nishijo M, Morikawa Y, et al. Increased urinary β2-microglobulin and mortality rate by cause of death in cadmium-polluted area. Environ Health Prev Med. 1996;1:144–8.

34. Arisawa K, Uemura H, Hiyoshi M, Dakeshita S, Kitayama A, Saito H, Soda M. Cause-specific mortality and cancer incidence rates in relation to urinary beta2-microglobulin: 23-year follow-up study in a cadmium-polluted area. Toxicol Lett. 2007;173(3):168–74.
35. Maruzeni S, Nishijo M, Nakamura K, et al. Mortality and causes of deaths of inhabitants with renal dysfunction induced by cadmium exposure of the polluted Jinzu River basin, Toyama, Japan; a 26-year follow-up. Environ Health. 2014;13:18. https://doi.org/10.1186/1476-069X-13-18.
36. Nakagawa H, Tabata M, Morikawa Y, et al. High mortality and shortened life-span in patients with *itai-itai* disease and subjects with suspected disease. Arch Environ Health. 1990;45:283–7.
37. Nishijo M, Nakagawa H, Suwazono Y, Nogawa K, Kido T. Causes of death in patients with *itai-itai* disease suffering from severe chronic cadmium poisoning: a nested case-control analysis of a follow-up study in Japan. BMJ Open. 2017;7(7):e015694.
38. Chen C, Xun P, Nishijo M, et al. Cadmium exposure and risk of lung cancer: a meta-analysis of cohort and case-control studies among general and occupational populations. J Expo Sci Environ Epidemiol. 2016a;26(5):437–44.
39. Chen C, Xun P, Nishijo M, et al. Cadmium exposure and risk of prostate cancer: a meta-analysis of cohort and case-control studies among the general and occupational populations. Sci Rep. 2016b;6:25814. https://doi.org/10.1038/srep25814.
40. Wu X, Zhu X, Xie M. Association between dietary cadmium exposure and breast cancer risk: an updated meta-analysis of observational studies. Med Sci Monit. 2015;21:769–75.
41. Sand S, Becker W. Assessment of dietary cadmium exposure in Sweden and population health concern including scenario analysis. Food ChemToxicol. 2012;50(3–4):536–44.
42. Arnich N, Sirot V, Rivière G, et al. Dietary exposure to trace elements and health risk assessment in the 2nd French Total Diet Study. Food Chem Toxicol. 2012;50(7):2432–49.
43. Kim SH, Lim YW, Park KS, et al. Relation of rice intake and biomarkers of cadmium for general population in Korea. J Trace Elem Med Biol. 2017;43:209–16. https://doi.org/10.1016/j.jtemb.2017.04.010.
44. Song Y, Wang Y, Mao W, et al. Dietary cadmium exposure assessment among the Chinese population. PLoS One. 2017;12(5):e0177978. https://doi.org/10.1371/journal.pone.0177978.
45. Watanabe T, Zhang ZW, Moon CS, et al. Cadmium exposure of women in general populations in Japan during 1991-1997 compared with 1977-1981. Int Arch Occup Environ Health. 2000;73(1):26–34.
46. Muramatsu S. Research about cadmium-pollution in the Kakehashi River basin. Jap J Pub Health (in Japanese). 1974;21:299–308.
47. Ikeda M, Ezaki T, Tsukahara T, Moriguchi J. Dietary cadmium intake in polluted and non-polluted areas in Japan in the past and in the present. Int Arch Occup Environ Health. 2004;77(4):227–34.
48. Nakashima K, Kobayashi E, Nogawa K, et al. Concentration of cadmium in rice and urinary indicators of renal dysfunction. Occup Environ Med. 1997;54(10):750–5.
49. Nogawa K, Ishizaki A. A comparison between cadmium in rice and renal effects among inhabitants of the Jinzu River basin. Environ Res. 1979;18(2):410–20.
50. Osawa T, Kobayashi E, Okubo Y, et al. A retrospective study on the relation between renal dysfunction and cadmium concentration in rice in individual hamlets in the Jinzu River basin, Toyama prefecture, Japan. Environ Res. 2001;86(1):51–9.
51. Akesson A, Julin B, Wolk A. Long-term dietary cadmium intake and postmenopausal endometrial cancer incidence: a population-based prospective cohort study. Cancer Res. 2008;68:6435–41.
52. Julin B, Wolk A, Bergkvist L, et al. Dietary cadmium exposure and risk of postmenopausal breast cancer: a population-based prospective cohort study. Cancer Res. 2012;72:1459–66.
53. Itoh H, Iwasaki M, Sawada N, et al. Dietary cadmium intake and breast cancer risk in Japanese women: a case-control study. Int J Hyg Environ Health. 2014;217(1):70–7.
54. Eriksen KT, Halkjær J, Meliker JR, et al. Dietary cadmium intake and risk of prostate cancer: a Danish prospective cohort study. BMC Cancer. 2015;15:177. https://doi.org/10.1186/s12885-015-1153-9.

55. Kjellström T, Nordberg GF. A kinetic model of cadmium metabolism in the human being. Environ Res. 1978;16:248–69.
56. Inskip H, Beral V, McDowall M. Mortality of Shipham residents: 40-year follow-up. Lancet. 1982;1:896–9.
57. Elliott P, Arnold R, Cockings S, et al. Risk of mortality, cancer incidence, and stroke in a population potentially exposed to cadmium. Occup Environ Med. 2000;57:94–7.
58. Eriksen KT, McElroy JA, Harrington JM, et al. Urinary cadmium and breast cancer: a prospective Danish cohort study. J Natl Cancer Inst. 2017;109(2):1–7. https://doi.org/10.1093/jnci/djw204.
59. Suwazono Y, Nogawa K, Morikawa Y, et al. Impact of urinary cadmium on mortality in the Japanese general population in cadmium non-polluted areas. Int J Hyg Environ Health. 2014;217(8):807–12.
60. Suwazono Y, Nogawa K, Morikawa Y, et al. All-cause mortality increased by environmental cadmium exposure in the Japanese general population in cadmium non-polluted areas. J Appl Toxicol. 2015a;35(7):817–23.
61. WHO/IPCS. Cadmium, environmental health criteria 134. Gingiva, Switzerland: World Health Organization, International Programme on Chemical Safety, WHO; 1992.
62. Jarup L, Alfven T, Persson B, et al. Cadmium may be a risk factor for osteoporosis. Occup Environ Med. 1998;55:435–9.
63. Peng Q, Bakulski KM, Nan B, et al. Cadmium and Alzheimer's disease mortality in U.S. adults: updated evidence with a urinary biomarker and extended follow-up time. Environ Res. 2017;157:44–51.
64. Suwazono Y, Nogawa K, Morikawa Y, et al. Renal tubular dysfunction increases mortality in the Japanese general population living in cadmium non-polluted areas. J Expo Sci Environ Epidemiol. 2015b;25(4):399–404.
65. Aoshima K. *Itai-itai* disease: renal tubular osteomalacia induced by environmental exposure to cadmium —— historical review and perspectives. Soil Sci Plant Nutri. 2016;62:319–26.
66. Friberg L, Piscator M, Nordberg G. The *itai-itai* disease. In: Friberg L, Piscator M, Nordberg G, editors. Cadmium in the environment. Ohio: CRC press; 1971. p. 111–4.
67. Nogawa K, Kido T. *Itai-itai* disease and health effects of cadmium. In: Chang LW, editor. Toxicology of metals. New York: CRC press; 1996. p. 353–69.
68. Kasuya M. Present status of *itai-itai* disease patients as of January 12, 2001. Kankyo Hoken Report. 2002;68:41–54.
69. Aoshima K. *Itai-itai* disease: cadmium-induced renal tubular osteomalacia. Nihon EisegakuZasshi. 2012;67:455–63.

Chapter 6
Cadmium Exposure and Its Effects on the Health Status of Rice Farmers in Akita Prefecture

Hyogo Horiguchi

Abstract Many areas polluted by cadmium (Cd) from mines and smelters exist in Akita Prefecture, and a large number of rice farmers in these areas have been exposed to high levels of Cd through the consumption of home-harvested rice. Rice farmers with Cd nephropathy, particularly elderly women, have been identified through health examinations performed in previous and recent studies. Although recent exposure to Cd among farmers in Cd-polluted areas has been decreasing due to countermeasures, the continuous monitoring of their health status and health care for patients with Cd nephropathy are necessary because of the long biological half-life of Cd. Intensive health examinations for inhabitants of these areas and hospital-based examinations for patients with renal dysfunction are currently being performed in Cd-polluted areas.

Keywords Cadmium · Nephropathy · Akita · Farmer · Rice · Health examination Mine · Smelter

6.1 Introduction

Several areas polluted by cadmium (Cd) from mines and smelters exist in Japan. The Jinzu River basin in Toyama Prefecture was the most heavily polluted area, with rice farmers being exposed to Cd through the consumption of locally harvested rice as well as Cd-contaminated river water [1]. Many patients with Cd nephropathy, a multiple proximal tubular dysfunction, have been identified in these areas, and of these, 200 were officially diagnosed by 2017 with *itai-itai* disease, the most severe outcome of the hazardous effects of Cd showing osteomalacia and renal anemia [2, 3]. The recognition of *itai-itai* disease in the Jinzu River basin in the 1970s

H. Horiguchi (✉)
Department of Hygiene, Kitasato University School of Medicine, Sagamihara, Kanagawa, Japan
e-mail: hhyogo@med.kitasato-u.ac.jp

S. Himeno, K. Aoshima (eds.), *Cadmium Toxicity*, Current Topics in Environmental Health and Preventive Medicine,
https://doi.org/10.1007/978-981-13-3630-0_6

led to an investigation of other Cd-polluted areas in Japan in an attempt to find similar patients. For example, a large number of patients with Cd nephropathy were detected in the Kakehashi River basin in Ishikawa Prefecture [4], Tsushima Island in Nagasaki Prefecture [5], and Kosaka town in Akita Prefecture [6] by health examinations performed on inhabitants.

Follow-up studies were conducted on inhabitants exposed to Cd in the Kakehashi River basin [7] and Tsushima Island [8], and the findings obtained revealed the prolonged adverse renal effects of Cd. Follow-up studies have not yet been performed on the inhabitants of Kosaka town; therefore, the health status of inhabitants exposed to Cd remains unknown. Akita Prefecture has many scattered Cd-polluted areas, the total width of which is the largest in Japan, as described below, and Kosaka town accounts for only a small area [9]. Therefore, conflicts over environmental, agricultural, and human damage caused by mine activity have frequently occurred, leading to investigations on adverse health effects among individuals living near these mines throughout Akita Prefecture in around 1970 [10]. However, follow-up health examinations to elucidate the effects of Cd on health have not been conducted. Therefore, Akita Prefecture is regarded as "the largest, but forgotten Cd-polluted area in Japan." This chapter focuses on Akita Prefecture with an overlook of the historical and current status of Cd pollution and its effects on the health of its inhabitants, particularly rice farmers.

6.2 An Outline of Cd-Polluted Areas in Akita Prefecture

Akita Prefecture is located in the northern part of Honshu, the mainland of Japan. The weather is relatively cold, with heavy snowfall in winter. The number of sunlight hours in winter is the smallest in all 47 prefectures of Japan. Its total area is the 6th largest in all prefectures, while its population is 38th according to the Japanese population statistics in 2017, indicating depopulation (45th in population density) (Table 6.1). Rice farming is one of the important industries in Akita Prefecture: the area of rice fields and rice yield in 2016 were both the third highest in all prefectures, according to statistics released from the Ministry of Agriculture, Forestry, and Fisheries (Table 6.1). Therefore, the number of rice farmers in Akita Prefecture is estimated be one of the highest in Japan.

Table 6.1 Profile of Akita Prefecture

Items	Data	Notes
Total area	11,637.52 km^2	
Estimated population (in 2017)	999,444	38th in Japan
Population density (in 2017)	85.9/km^2	45th in Japan
Area of rice fields (in 2016)	124,700 ha	3rd in Japan
Rice yield (in 2016)	499,700 t	3rd in Japan

Akita Prefecture was previously rich in mineral resources, such as zinc, copper, lead, silver, and gold, which made it one of the most flourishing prefectures with a mineral industry in Japan. There have been up to 125 small and large non-iron mines in Akita Prefecture in the past few hundred years [9], although most of them are no longer operational due to the depletion of their resources. Since ores such as zinc, copper, and lead often contain Cd at relatively high concentrations(1:100–1:1000) [11], waste fluid from mines or emissions from smelters were contaminated with Cd, leading to the large distribution of Cd to the surrounding neighborhoods, including rice fields. Rice intensively absorbs Cd from soil, leading to the prominent accumulation of Cd in rice produced in Cd-polluted areas. According to an official announcement by the Ministry of the Environment of Japan, the cumulative width of "areas in which countermeasures for Cd pollution were required" formally designated by the Anti-Farm Soil Pollution Law based on the production of rice with a Cd concentration over the safety standard level was the highest in Akita Prefecture (Fig. 6.1). It is important to note that designated Cd-polluted areas in Akita Prefecture are even larger than those in Toyama Prefecture, in which a large number of patients have been diagnosed with *itai-itai* disease. The number of designated Cd-polluted areas in Akita Prefecture is 25, which is markedly higher than those among all prefectures, accounting for approximately 40% of the 64 designated areas (Fig. 6.2). These findings indicate that Cd-polluted areas of various sizes are scattered around Akita Prefecture, which is in contrast to the Jinzu River basin in Toyama Prefecture, in which heavily polluted areas are limited to one location.

The distribution of designated Cd-polluted areas in Akita Prefecture shows that they may be divided into three parts: northern, central, and southern (Fig. 6.3). These parts correspond to the locations of mines or smelters that were previously operational. In the northern part, there was a large-scale mine along with several satellite mines in Kosaka town, as described above. There was another large mine in the adjacent city, Kazuno, named the Osarizawamine. These mines had smelters, from which emissions caused severe damage to the surrounding woods. Although

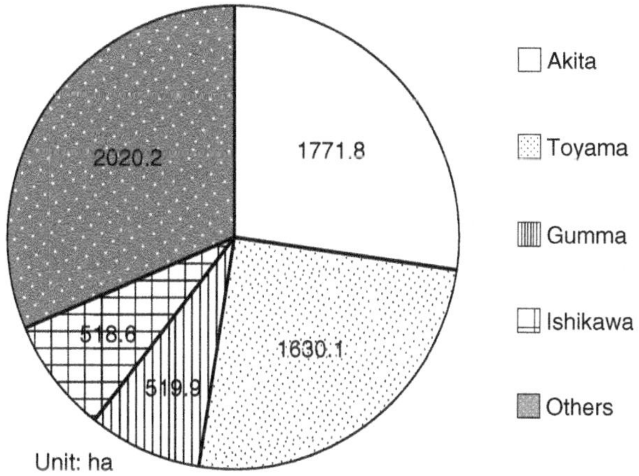

Fig. 6.1 Cadmium-polluted areas by prefecture in Japan, designated by the Anti-Farm Soil Pollution Law in 2015. The Ministry of the Environment of Japan

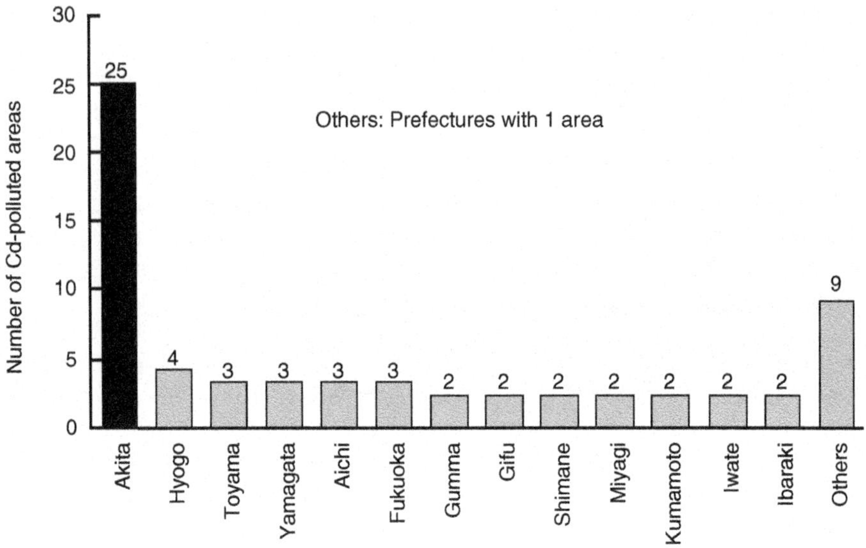

Fig. 6.2 Numbers of cadmium-polluted areas by prefecture in Japan, designated by the Anti-Farm Soil Pollution Law in 2015. The Ministry of the Environment of Japan

Fig. 6.3 A map of cadmium-polluted areas in Akita Prefecture designated by the Anti-Farm Soil Pollution Law in 2015. Gray circles indicate the designated areas, and open large circles indicate the northern, central, and southern parts of Cd-polluted areas. The Ministry of the Environment of Japan

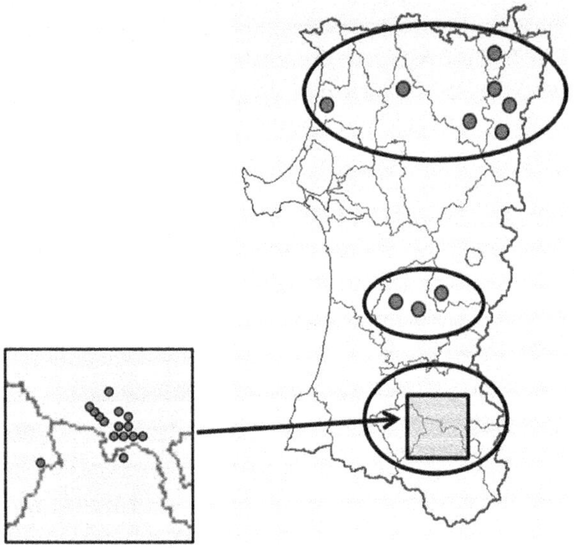

these two mines were located in different areas, a river from each mine merged into Yoneshiro River, along which there were also a few small mines. These three rivers were expected to be contaminated by Cd from these mines because designated Cd-polluted areas were scattered along these rivers, from which water was used to irrigate rice fields. In addition, particulate emissions from local smelting factories further increased Cd pollution.

The representative large mine in the central part, named the Arakawa mine, was operated in Kyowa town, and Arakawa River runs from this mine, which had a smelter. Effluents and emissions from the mine and smelter caused fishery and agricultural damage in downstream areas. Since the designated Cd-polluted areas were located along the river, rice fields downstream of the river were estimated to be polluted by Cd from the mine. There was a large mine, named the Yoshinomine, in the southern part; however, it did not have a smelter. Naruse River runs from this mine, and water was used for the irrigation of downstream rice fields. There were many designated Cd-polluted areas in a limited, concentrated manner, which corresponded to the areas of irrigation using river water potentially contaminated with Cd from the mine.

6.3 Health Examinations Previously Conducted in Akita Prefecture

Although detailed information is not available, investigations on the health status of inhabitants were sporadically performed in areas in which serious environmental damage was caused by the activities of mines or smelters in Akita Prefecture from around 1965 [10]. Systematic health examinations to elucidate the adverse effects of Cd among inhabitants aged 20 years or older in Cd-polluted areas throughout Akita Prefecture, including the northern, central, and southern parts, were initiated in 1972. These health examinations consisted of three steps: the first step included basic urinary tests for urinary protein and glucose levels as well as medical interviews, the second assessed urinary Cd levels and detected urinary tubular proteins, and the third involved careful physical examinations by hospitalization, such as X-rays of bones and assessments of serum alkaline phosphatase and urinary mineral levels. According to the records of examinations performed in 1972–1973, 3183 inhabitants in 15 Cd-polluted hamlets and 210 in a control hamlet, in which Cd concentrations in soil were 2.5 ppm (a geometric mean of 15 data sets) and 0.3 ppm, respectively, and Cd concentrations in rice were 0.52 ppm (a geometric mean of 15 data) and 0.13 ppm, respectively, were investigated. A total of 13% and 4.8% of the examinees in Cd-polluted and control areas, respectively, showed abnormal urinary protein and glucose levels, while 4% and 0.5% of those in Cd-polluted and control areas, respectively, showed a urinary Cd concentration of more than 10 μg/L. There were 22 examinees with urinary proteins of the tubular pattern, suggesting Cd nephropathy, in Cd-polluted areas, in contrast to none in the control area. Among them, 13 of the suspected examinees with Cd nephropathy were from Kosaka town in the northern Cd-polluted part, indicating that this area was the most severely affected in Akita Prefecture.

Two more detailed investigations were performed on the inhabitants of Kosaka town. In 1972–1975, health examinations were conducted by a research group at

Tohoku University on 147 men and women older than 35 years to older than 70 years in a hamlet adjacent to a large smelter [6]. The mean concentration of Cd in 77 samples of unpolished rice was 0.66 ± 0.72 μg/g. Thirty-three examinees exhibited renal glycosuria, tubular proteinuria, and aminoaciduria. The urinary concentration of β_2-microglobulin, a sensitive indicator of renal tubular dysfunction, in the examinees increased with age. The concentration of Cd in rice multiplied by their resident years closely correlated with the concentration of urinary β_2-microglobulin. Ten examinees were clinically diagnosed with multiple tubular dysfunctions, including one that had a suspected sign of osteomalacia.

Another study was performed by the Akita Prefecture Institute of Public Health in 7 hamlets located at a shortest distance from the mine and smelter in 1975, in which 156 men and women aged 50–69 years were investigated, along with 93 farmers in a control town located approximately 70 km southwest of the Cd-polluted area [12, 13]. The geometric averages of fecal Cd contents, which are good estimators of daily Cd intake, in the inhabitants of Kosaka town and the control were 146 μg/day and 41 μg/day, respectively, and geometric averages of urinary Cd concentrations were 7.5 μg/L and 2.1 μg/L, respectively. The prevalence of tubular proteinuria, defined as an increased urinary β_2-microglobulin concentration (>700 μg/L), was 14% in Kosaka town and 3% in the control, showing an increase with age and exposure duration. In 1976, a follow-up health examination was conducted by the same institute as a project commissioned by the government on 427 men and women in 9 hamlets of Kosaka town and 403 in another control, showing a significant decrease in TRP (tubular reabsorption of phosphate), which indicates Cd nephropathy, among examinees older than 70 years [14].

These findings indicated that adverse health effects were prominent among inhabitants in Cd-polluted areas in Akita Prefecture when mines and smelters were actively operated, particularly in Kosaka town of the northern part. However, health examinations on Cd-exposed inhabitants do not appear to have been followed up for approximately 30 years. During that time, mines and smelters ceased their operations because of the depletion of their mineral resources, and local environmental Cd pollution appeared to have decreased. In addition, the clinical manifestations of patients with Cd nephropathy were too weak to be recognized in an ordinary medical examination. Therefore, Cd-polluted areas in Akita Prefecture appear to have been largely forgotten.

6.4 A Recent Study in Akita Prefecture

In 2001–2004, health examinations were performed on female rice farmers aged 40 years or older in the northern part of Cd-polluted areas in Akita Prefecture by a research group at Jichi Medical University [15, 16]. The areas investigated included

Odate City, through which Yoneshiro River runs from the two large mines, as well as Kosaka town and Kazuno City, near to which large mines and smelters operated. The main purpose of this study was to evaluate the safety standard of Cd concentrations in rice, which was set as 0.4 μg/g according to a notification made by a chief officer of the Japanese Food Agency in 1983. Since then, this safety standard has been applied to rice on the market for general consumers. However, farmers in Cd-polluted areas were at risk of excessive exposure to Cd because they were inclined to consume home-harvested rice that may have been over the standard. Approximately 10% of home-harvested rice provided by the examinees showed Cd concentrations over the standard.

In this study, examinees were divided into two groups, the Kosaka and Kazuno area, near to large mines and smelters, and the Odate area, located downstream from them, consisting of 623 and 355 female farmers, respectively, and were compared with 222 female farmers in a control area located approximately 50 km west of Cd-polluted areas. Blood and urinary Cd levels were higher in examinees in Cd-polluted areas than in the control, showing age-dependent increases. Furthermore, these levels were higher in the Kosaka and Kazuno area than in the Odate area, reflecting their geographical relationship with Cd pollution. For example, the medians of urinary Cd levels in examinees aged 70 years or older were 2.99, 4.90, and 9.34 μg/g cr in the control, Odate area, and Kosaka and Kazuno area, respectively. It is important to note that urinary Cd levels in elderly examinees of the Kosaka and Kazuno area were close to the threshold to induce Cd nephropathy, namely, 10 μg/g cr [17]. On the other hand, no significant differences were observed in urinary β_2-microglobulin levels between the control and Odate area, and only examinees aged 70 years or older in the Kosaka and Kazuno area showed a significantly higher level of urinary β_2-microglobulin than the control. In addition, a patient with Cd nephropathy had a urinary β_2-microglobulin level higher than 10,000 μg/g cr [18] among the examinees aged 70 years or older in the Kosaka and Kazuno area. These findings indicate that Cd exposure and its adverse renal effects among rice farmers have persisted for decades since the previous studies described above were performed, and this may be attributed to the continuous consumption of self harvested rice with high Cd concentrations as well as the extremely long biological half-life of Cd in the body of 10–30 years [11].

6.5 Current Situation in Cd-Polluted Areas in Akita Prefecture

Along with the restoration of Cd-polluted paddy fields, the flooding of paddy fields, through which rice is cultivated before and after heading during August to lower Cd absorption by rice from the soil [19], was effectively adopted in Cd-polluted areas in Akita Prefecture after 2000. Due to these countermeasures, rice with a Cd concentration over the safety standard has rarely been detected. Therefore, it is highly plausible that recent Cd exposure levels through the consumption of self-harvested rice among farmers in Cd-polluted areas have been declining. However, because of the long biological half-life of Cd, the renal tubular function of farmers previously exposed to an excessively high level of Cd may be affected with aging, even after a decrease in Cd exposure.

Therefore, a research group from Akita University and Kitasato University started conducting intensive health examinations in 2009 with the aim of clarifying the actual state of whole Cd-polluted areas in Akita Prefecture and contributing to the health status of farmers, such as the prevention of future Cd exposure among younger farmers and providing health care for individuals exhibiting adverse Cd effects. Health examinations are available for all rice farmers, both male and female, aged 40 years or older, in every hamlet in Cd-polluted areas. They are ongoing in 2017, having covered mainly the northern part, with a more than 70% response rate on average; investigations on the central and southern parts will be conducted in the future. Preliminary data from more than 1000 examinees including a control area shows higher average levels of blood and urinary Cd in addition to approximately 10 patients with Cd nephropathy in hamlets around the mines and smelters (unpublished).

However, the approach of health examinations, to which only people in good condition may go, does not account for infirm individuals who cannot participate and may have severe Cd nephropathy or even *itai-itai* disease. Therefore, in parallel with health examinations, the same research group is conducting Cd-related examinations on patients with renal dysfunction who attend or are admitted to local core hospitals or clinics in Cd-polluted areas. This aim of this hospital-based approach is to detect patients with Cd nephropathy by measuring blood and urinary Cd and urinary β_2-microglobulin levels. Preliminary data supports the effectiveness of this approach, with several elderly patients with Cd nephropathy being identified to date (unpublished). Therefore, these approaches to inhabitants in Cd-polluted areas function in a complementary manner to detect patients with Cd adverse effects.

6.6 Conclusion

In Akita Prefecture, there are still a large number of rice farmers who have been exposed to high levels of Cd through the consumption of home-harvested rice produced in Cd-polluted areas, even after these mines and smelters ceased their operations. Some elderly women have developed Cd nephropathy, although the number detected to date is small. Therefore, the actual status of Cd exposure among rice farmers in all Cd-polluted areas needs to be clarified and preventive measures and health care provided. Local health examinations for inhabitants and a hospital-based approach to patients with renal dysfunction are currently underway in Cd-polluted areas.

References

1. Aoshima K. Epidemiology of renal tubular dysfunction in the inhabitants of a cadmium-polluted area in the Jinzu River basin in Toyama prefecture. Tohoku J Exp Med. 1987;152:151–72.
2. Tsuchiya K. Causation of Ouch-Ouch Disease (Itai-ItaiByõ)--an introductory review. I. Nature of the disease. Keio J Med. 1969;18:181–94.
3. Kasuya M, Teranishi H, Aoshima K, Katoh T, Horiguchi H, Morikawa Y, Nishijo M, Iwata K. Water pollution by cadmium and the onset of Itai-itai disease. Wat Sci Tech. 1992;25:149–56.
4. Nogawa K, Ishizaki A, Kawano S. Statistical observations of the dose-response relationships of cadmium based on epidemiological studies in the Kakehashi River basin. Environ Res. 1978;15:185–98.
5. Iwata K, Saito H, Moriyama M, Nakano A. Renal tubular function after reduction of environmental cadmium exposure: a ten-year follow-up. Arch Environ Health. 1993;48:157–63.
6. Saito H, Shioji R, Hurukawa Y, Nagai K, Arikawa T. Cadmium-induced proximal tubular dysfunction in a cadmium-polluted area. Contrib Nephrol. 1977;6:1–12.
7. DucPhuc H, Kido T, Dung Manh H, Thai Anh L, Phuong Oanh NT, Okamoto R, Ichimori A, Nogawa K, Suwazono Y, Nakagawa H. A 28-year observational study of urinary cadmium and β_2-microglobulin concentrations in inhabitants in cadmium-polluted areas in Japan. J Appl Toxicol. 2016;36:1622–8.
8. Liu XJ, Arisawa K, Nakano A, Saito H, Takahashi T, Kosaka A. Significance of cadmium concentrations in blood and hair as an indicator of dose 15 years after the reduction of environmental exposure to cadmium. Toxicol Lett. 2001;123:135–41.
9. Akita prefecture underground resources development promotion association, Foundation of Akita mining industry association. Akita-ken Kohzanshi (in Japanese), or "Historical records of mines in Akita prefecture"; 2005.
10. Kamiyama S. Health examinations for inhabitants in cadmium-polluted areas in Akita prefecture (in the years of 1972-1973). In: Kankyo Hoken Report No. 36. Tokyo: Japanese Association of Public Health; 1976. p. 124–5. (in Japanese).
11. Nordberg GF, Nogawa K, Nordberg M. Cadmium. In: Nordberg GF, Fowler BA, Nordberg M, editors. Handbook on the toxicology of metals. 4th ed. Burlington: Academic Press; 2015. p. 667–716.
12. Kojima S, Haga Y, Kurihara T, Yamawaki T. A comparison between fecal cadmium and urinary $beta_2$-microglobulin, total protein, and cadmium among Japanese farmers. An epidemiological study of cooperation between Japan and Sweden. Environ Res. 1977;14:436–51.
13. Haga Y, Kurihara T,Yamawaki T, Kojima S. Research report on environmental health studies of heavy metal pollution: No. 3, In Report of The Akita Pref. Institute of Public Health No. 19, Akita; 1975, p. 69–75 (in Japanese).
14. Haga Y, Kobayashi T,Miura H, Kato A. Research report on environmental health studies of heavy metal pollution: No. 5, In Report of The Akita Pref. Institute of Public Health No. 21, Akita; 1977, p. 101–109 (in Japanese).
15. Horiguchi H, Oguma E, Sasaki S, Miyamoto K, Ikeda Y, Machida M, Kayama F. Dietary exposure to cadmium at close to the current provisional tolerable weekly intake does not affect renal function among female Japanese farmers. Environ Res. 2004;95:20–31.
16. Horiguchi H, Oguma E, Sasaki S, Okubo H, Murakami K, Miyamoto K, Hosoi Y, Murata K, Kayama F. Age-relevant renal effects of cadmium exposure through consumption of home-harvested rice in female Japanese farmers. Environ Int. 2013;56:1–9.
17. Bernard A. Confusion about cadmium risks: the unrecognized limitations of an extrapolated paradigm. Environ Health Perspect. 2016;124:1–5.
18. Bernard A. Renal dysfunction induced by cadmium: biomarkers of critical effects. Biometals. 2004;17:519–23.
19. Arao T, Ishikawa S, Murakami M, Abe K, Maejima Y, Makino T. Heavy metal contamination of agricultural soil and countermeasures in Japan. Paddy Water Environ. 2010;8:247–57.

Chapter 7
Dietary Cadmium Intake Among General Populations in Japan and East Asia

Masayuki Ikeda, Takao Watanabe, Haruo Nakatsuka, Shinichiro Shimbo, Chan-Seok Moon, Hyeran Yang, Peiyu Wang, and Zuowen Zhang

Abstract Background: Cadmium (Cd) is a ubiquitous environmental pollutant that has insidious toxicities after chronic exposures. Daily foods are almost exclusive exposure sources for general populations, and the staple food of paddy rice plays an important role as the exposure source. **Study Aim:** The aims are to elucidate current levels and future trends of dietary Cd intake (Cd-D) in Japan, Korea,

M. Ikeda (✉)
Kyoto University, Kyoto, Kyoto, Japan

Tohoku University, Sendai, Miyagi, Japan

Kyoto Industrial Health Association, Kyoto, Kyoto, Japan
e-mail: m-71-ikeda@nifty.com

T. Watanabe
Tohoku Bunkyo University, Yamagata, Yamagata, Japan

H. Nakatsuka
Kanazawa Gakuin College, Kanazawa, Ishikawa, Japan

S. Shimbo
Kyoto Women's University, Kyoto, Kyoto, Japan

C.-S. Moon
Catholic University of Pusan, Busan, South Korea

H. Yang
Seoul Metropolitan Government Research Institute of Public Health and Environment, Seoul, South Korea

P. Wang
Peking University, Beijing, China

Z. Zhang
National Natural Science Foundation, Beijing, China

S. Himeno, K. Aoshima (eds.), *Cadmium Toxicity*, Current Topics in Environmental Health and Preventive Medicine,
https://doi.org/10.1007/978-981-13-3630-0_7

and China. **Materials and Methods:** Publications since 1960 on Cd-D of adult populations in areas with no known Cd pollution were retrieved; in practice 19, 14, and 18 reports were available for Japan, Korea, and China, respectively. Food consumption data were primarily based on food duplicate collection or market basket methods. Flame/flameless atomic absorption spectrometry (AAS) and inductively coupled plasma-mass spectrometry (ICP-MS) were used for Cd analysis. **Results and Discussion:** In Japan, the earliest reports on Cd-D in the late 1960s gave Cd-D values of 41–116 μg/day. Surveys in 1977–1981 resulted in 37 μg/day, and the estimates for 2003–2007 were 11.5–16.5 μg/day. In Korea, Cd-D of 71 μg/day was first reported in 1980. Cd-D of >20 μg/day followed in the 1990s, 10–20 μg/day in the 2000s, and 5–7 μg/day in 2011. The reduction in Japan and Korea was achieved by regulatory and technical efforts to remove high-Cd rice from markets. Unexpected reduction in rice consumption was also observed. In China, the first report on Cd-D gave 13.8 μg/day for the year 1990, and the latest publication in 2012 reported a Cd-D of 15.5 μg/day. However, there was a wide variation in Cd-D up to 41.8 μg/day during this period, which made the trend complex. The time span (1990 to 2015) may be too short for trend analysis. The study numbers may be insufficient for the vast country. **Conclusion:** Analyses for Japan and Korea revealed that Cd-D has been gradually decreasing to the level well below 20 μg/day or even lower. Further follow-up studies are desirable for the analysis in case of China.

Keywords Adult population · Cadmium · China · Dietary intake · Japan · Korea

7.1 Introduction

It is known that cadmium (Cd) is a ubiquitous environmental pollutant, known to induce renal tubular dysfunction and then bone metabolism disturbance after a long-term exposure [1–3]. The outbreak of the *itai-itai* disease in the Jinzu River basin [4, 5] (for details including clinical picture, see Chaps. 1 and 2) caused serious concern of the public in Japan to Cd, the causative agent.

Cd intake through daily foods (Cd-D) is the major source of the exposure for general population [6]. It became further clear that Cd exposure is primarily through rice [5]. Paddy rice is an important source of energy for daily life, not only in Japan [7] but also in Korea and in the southern part of China. Thus, the concern has been inevitably expanded over East Asian countries [8–14]. Attention in Japan then has been shifting from the existing level of exposure to the procedures and resulting effects to reduce Cd-D (see Chap. 10 for details).

This chapter is focused on the annual changes in Cd-D in these three East Asian countries. Previous surveys have disclosed that Cd-D levels were low in South-East Asia [6]. Reviews on Cd-D in Japan have been previously published elsewhere [15–17].

7.2 Materials and Methods

Publications were retrieved for Cd-D in Japan, Korea, and China from 1969 when Cd-D in the *itai-itai* disease-endemic area was officially surveyed for the first time (for details, see Sect. 7.3.1 and Chaps. 1 and 2) till the end of March, 2017. In practice, 19, 14, and 18 sets of data in 10, 7, and 12 publications, respectively, on Cd-D were available (see Sect. 7.3 for details).

For food intake survey [18], two methods were prevalently used; one was the 24-hours food duplicate sampling (i.e., total diet method) [19, 20], and the other was the market basket method [21, 22]. The memory recall method was also used by several authors [23, 24]. Some studies [12, 25] employed the results of national survey on food intake as the database for food intake.

Regarding instrumental analysis of the food materials for Cd, the flame or graphite furnace (GF; flameless) atomic absorption spectrometry (AAS) was employed in early time studies [26], and the use of inductively coupled plasma-mass spectrometry (ICP-MS) became prevalent in recent studies [23, 27, 28]. Tsuchiya and Iwao [5] confirmed that the results by the flame AAS and by GFAAS were compatible. Compatibility of GFAAS and ICP-MS in analysis results has also been confirmed [29, 30].

As Cd-D is known to be distributed log-normally, geometric means (GMs) and geometric standard deviations (GSDs) were taken as representative parameters of distribution. In case original results were given as arithmetic means (AMs) and arithmetic standard deviations (ASDs), GMs and GSDs were estimated from the AMs and ASDs by use of the moment method [31].

Some data were presented in terms of Cd-D in μg/kg body weight/day, possibly to evaluate in comparison with PTMI (provisional tolerable monthly intake; for details, see Sect. 7.4). A body weight of 59.35 kg was considered for adult populations in Japan [7] (65.8 kg for men and 52.9 kg for women), 63.9 kg for adult populations in Korea [32] (71.0 kg for men and 56.8 kg for women), and 61.75 kg for adult populations in China [33] (66.2 kg for men and 57.3 kg for women).

Modified exponential curves were drawn by use of software, SPSS (Version 20; IBM, Armonk NY, USA) and Excel Tokei 2012 (Version 1.15; Social Survey Research Information, Ltd., Tokyo, Japan), respectively.

7.3 Results

7.3.1 Observation in Japan

In Japan, 19 Cd-D values (as group means) were available (Table 7.1). The earliest surveys on Cd-D in 1969 to compare with the exposure level among the *itai-itai* disease victims (which was as high as 600 μg/day or even higher [34, 35]) reported

Table 7.1 Dietary cadmium intake (Cd-D) in Japan*

Authors and year of publication	Ref. No.	Survey year	Methods[a]	No. of samples	Cd-D (μg/day)
Environmental Agency 1972; Kitamura 1972	[34, 35]	1968	MB	uk[b]	60[c]
Fukushima 1972	[4]	Ibid.	MB	uk[b]	47.0[c]
Japan Public Health Association, 1970	[36]	1969	FD	3	115.5[M,d]
Ibid.		Ibid.	FD	3	80.8[W,d]
Ibid.		Ibid.	FD	3	95.7[M,d]
Ibid.		Ibid.	FD	3	94.4[W,d]
Ibid.		Ibid.	FD	3	55.2[M,d]
Ibid.		Ibid.	FD	3	55.4[W,d]
Yamagata and Iwashima 1975	[26]	1972	FD	(28 sites)	23.2[e]
Saito 2003	[37]	1976	FD	10	50.4[f]
Tsuchiya and Iwao 1978	[5]	1978	FD	10	60.1[M,d]
Ibid.		Ibid.	FD	10	39.3[M,d]
Ibid.		Ibid.	FD	10	58.9[M,d]
Watanabe et al. 1985	[19]	1977–1981	FD	368	43.9[M]
Ibid.		Ibid.	FD	674	37.0[W]
Saito 2003	[37]	1983	FD	6	28.8[f]
Ikeda et al. 2000	[38]	1991–1997	FD	607	25.5[W]
Ikeda et al. 2015	[39]	Ibid.		(30 sites)[g]	16.5[h]
Ibid.		Ibid.		(30 sites)[g]	11.5[i]

[a]*MB* market basket method, *FD* food duplicate
[b]*uk* unknown, assumedly MB
[c]Only one value is given
[d]Recalculated from published data
[e]Calculated from AM and ASD
[f]Possibly AM
[g]Number of sites to establish the estimation equation
[h]Estimated from Cd-B
[i]Estimated from Cd-Ucr
[M]Value for men
[W]Value for women
*This table was reproduced (after revision) from Table 7.1 in M. Ikeda et al. Int Arch Occup Environ Health 2004; 77: 227–234. Permission for reproduction was granted by Springer Verlag

a Cd-D of 60 μg/day [34, 35], whereas a slightly lower value of 47.0 μg/day was also reported [21]. Although the control surveys were conducted in six nonpolluted areas, the Cd-D levels reported were in the ranges of 55.2 to 115.5 μg/day [36]. It should be noted however that the number of cases was only three per sex and per site (thus 18 in total). The survey by Yamagata and Iwashima [26] gave a lower Cd-D of 31.3 μg/day as AM (28.3 μg/day as ASD) for general Japanese populations. The GM would be 23.2 μg/day when the moment method [31] is applied. In repeated surveys in the same control areas, Saito [37] observed substantial reduction (by 43%) in Cd-D from 50.4 μg/day in 1976 to 28.8 μg/day in 1983.

Tsuchiya and Iwao [5] observed Cd-D of 39 to 60 μg/day in their studies in areas selected as controls to three Cd-polluted areas. Kitamura [35] estimated that Cd from rice was 23 μg/day (38%) out of 60 μg/day for total foods. Tsuchiya and Iwao [5] also identified rice as the leading source of dietary Cd burden.

Watanabe et al. [19] conducted a nationwide survey in 49 sites all over Japan (from Hokkaido Island in the north to Okinawa Islands in the south) in 1977 to 1981. The Cd-D was 43.9 μg/day for 368 men and 37.0 μg/day for 674 women, respectively. The Cd-D was higher for men than for women possibly because men took more rice than women. In a nationwide follow-up survey on 607 women in 30 sites, Ikeda et al. [38] observed a decrease in Cd-D to 25.5 μg/day, the decrease being by 31%. Since then, there had been no survey on Cd-D in Japan. In the place, however, Ikeda et al. [39] conducted a whole country survey on Cd in blood (Cd-B) and Cd in urine (as corrected for creatinine; Cd-U_{cr}). It became clear that Cd-D correlates significantly with Cd-B and also with Cd-U_{cr} ($p < 0.01$ for both [39]). The correlation equations were $Z = -1.13 + 14.36X$ and $Z = 5.35 + 44.90Y$, where X = Cd-B (μg/L), Y = Cd-U_{cr} (μg/g cr), and Z = Cd-D (μg/day). Taking advantage of the equations, the Cd-D was estimated from Cd-B and Cd-U as 16.5 and 11.5 μg/day, respectively. Although the two estimates differ from each other by 30%, both values together suggest Cd-D of well less than 20 μg/day.

7.3.2 *Observation in Korea*

In Korea, 14 Cd-D data were available (Table 7.2) since 1980 when Yum et al. [8] reported an estimate of 70.5 μg/day. This value is the highest ever reported for Korea. In addition, Yum et al. [8] noted an increase in Cd in rice from <0.035 mg/kg in 1970 to 0.11 mg/kg in 1980. No survey had been made till mid-1990s when Moon et al. [10] reported Cd-D in a range of 14.3 to 24.2 μg/day. The surveys were conducted in two large cities, one smaller city, and a rural village. Urban-rural difference in Cd-D was not remarkable.

In the 2000s, Moon et al. [40] surveyed Cd-D of 38 pairs of mothers and children. Cd-D for the mothers was 16.7 μg/day. The Cd-D for children, 0.457 μg/kg body weight/day, was higher than the value for the mothers, 0.300 μg/kg body weight/day, possibly because children needed more foods (per body weight) for their growth in addition to daily activities. In the 2000s, Oh et al. [41] published a substantially lower Cd-D estimate of 12.61 μg/day. Oh et al. [41], in addition, confirmed that foods were almost an exclusive source of Cd for general population in Korea. In 2012, Moon et al. [25, 42] reported that Cd-D has been decreasing in Korea; estimated Cd-D [25] was 17.02, 16.57, 16.08, and 14.82 μg/day for 1998, 2001, 2005, and 2007, respectively. In 2010 to 2011, Lim et al. [28] conducted a large-scale survey covering the whole country and observed a Cd-D of 0.18 μg/kg body weight/day. The value is equivalent to 11.5 μg/day when 63.9 kg was assumed

Table 7.2 Dietary cadmium intake (Cd-D) in Korea*

Authors and year of publication	Ref. No.	Survey year	Methods[a]	No. of samples	Cd-D (μg/day)
Yum et al. 1980	[8]	1980	MB	nk[b]	70.53
Moon et al. 1995	[10]	1994	FD	24	14.3[W]
Ibid.		1994	FD	47	24.3[W]
Ibid.		1994	FD	29	19.6[W]
Ibid.		1994	FD	41	24.2[W]
Moon et al. 2012	[25]	1998	NHNES[c]	nk[b]	17.02[W]
Moon et al. 2003	[40]	2000	FD	38	16.7[W]
Moon et al. 2012	[25]	2001	NHNES[c]	nk[b]	16.57[W]
Oh et al. 2006	[4]	2001	MB	30	12.61
Moon et al. 2012	[25]	2005	NHNES[c]	nk[b]	16.08[W]
Ibid.		2007	NHNES[c]	nk[b]	14.82[W]
Lim et al. 2015	[28]	2010–2011	MB	4876[d]	11.5[e,f]
Huang et al. 2011	[43]	2011[g]	MR	239	6.53[M,h]
Ibid.		2011[g]	MR	404	5.36[W,h]

[a]*MB* market basket method, *FD* food duplicate, *MR* memory recall
[b]*uk* unknown
[c]National Health and Nutrition Examination Survey, Korea (2005, 2008)
[d]43.5% were adults
[e]Median
[f]Taking 63.9 kg body weight for adult Korean population [32]
[g]Publication year in place of survey year
[h]Estimate from AM and ASD
[W]Value for women
[M]Value for men
*This table was reproduced (after revision) from Table 7.1 in C.-S. Moon et al. Environ Health Prev Med 2016; 21: 118–128. Permission for reproduction was granted by The Japanese Society for Hygiene

as an average body weight [32]. About the same time, Huang et al. [43] estimated that Cd-D was 6.53 and 5.36 μg/day for men and women, respectively, in a survey in six sites in two provinces.

7.3.3 *Observation in China*

In China, 18 Cd-D data were available on Cd-D (Table 7.3). Gao et al. [44] reported that Cd-D was 13.8, 19.4, and 22.2 μg/day for 1990, 1992, and 2000, respectively (assumedly AM values), with an apparent trend of year-dependent increase. Nevertheless, four succeeding studies gave substantially lower values for Cd-D. Zhang et al. [45] reported 6.3 and 6.1 μg/day for two metropolitan cities, and Ikeda et al. [6] gave 8.7 μg/day as an average for four cities in the continent.

Table 7.3 Dietary cadmium intake (Cd-D) in China*

Authors and year of publication	Ref. No.	Survey year	Methods[b]	No. of samples	Cd-D (μg/day)
Gao et al. 2006	[44]	1990	FD	uk[c]	13.8[f]
Ibid.		1992	FD	uk[c]	19.4[f]
Zhang et al. 1997	[45]	1993–1995	FD	50	6.3[w]
Ibid.		1993–1995	FD	50	6.1[w]
Ikeda et al. 2000	[6]	1991–1998	FD	200	8.7[w]
Watanabe et al. 1998	[46]	1996	FD	50	6.43
Ibid.		1996	FD	50	5.93
Watanabe et al. 2000	[47]	1997	FD	49	5.83
Ibid.		1997	FD	49	6.79
Ibid.		1997	FD	50	5.64
Gao et al. 2006	[44]	2000	FD	uk[c]	22.2[f]
Yuan et al. 2014	[12]	2002	MB	CNNHS[d]	27.5[f]
Zheng et al. 2007	[22]	2002	MB	CNNHS[d]	41.80[f]
Wang et al. 2012	[27]	2012[a]	FD	120	15.5[M]
Tang et al. 2014	[49]	2008	MB	na[e]	15.42[f]
He et al. 2013	[50]	2008	MR	207	9.63[g]
Chen et al. 2014	[23]	2014[a]	MR	5008	17.1[f,h]
Wang et al. 2015	[24]	2015[a]	FD	103	20.4[f,h]

[a]Publication year in place of survey year
[b]*MB* Market basket, *FD* food duplicate, *MR* memory recall, *uk* unknown
[c]*uk* unknown
[d]Data cited from China National Nutrition and Health Survey [48]
[e]*na* not available
[f]Assumedly AM
[g]Median
[h]A body weight of 61.75 kg was assumed for men and women in combination
[M]Value for men
[W]Value for women
*This table was reproduced (after revision) from Table 7.5 in T. Watanabe et al. Biomed Res Trace Elem 2017; 28: 18–27. Permission for reproduction was granted by Japanese Society for Biomedical Research on Trace Elements

Watanabe et al. [46] studied one provincial capital and one rural village to find Cd-D values of 6.43 and 5.93 μg/day, respectively. Watanabe et al. [47] then studied another provincial capital and two farming villages where Cd-D was 5.83, 6.79, and 5.64 μg/day, respectively. Thus, Cd-D values reported were well less than 10 μg/day in these studies.

Yuan et al. [12] took advantage of national data for food consumption [48] and made a large-scale survey for Cd contents in various food materials. The calculation through combination of two databases gave Cd-D of 3.12 μg/kg body weight/week, which is equivalent to 27.5 μg/day when 61.75 kg was assumed to be an average body weight [33]. Zheng et al. [22] made a similar calculation for a local city to find a Cd-D of 41.8 g/day.

Wang et al. [27] collected 120 food duplicate samples from adult residents in four cities and analyzed for various elements including Cd. The Cd-D was 15.5 μg/day as GM. Tang et al. [49]) used a locally produced food consumption database and established a mineral content list for locally available food materials. By use of the two databases, a Cd-D of 15.42 μg/day was obtained. By use of food frequency questionnaire, He et al. [50] studied Cd-D for citizens (at the ages of 40 years and over) in a metropolis. The Cd-D thus estimated was 9.63 μg/day for non-smokers and 18.24 μg/day for total participants. The difference might suggest an additional smoking-associated Cd burden. Chen et al. [23] used existing food consumption data (the data were collected by 24-memory recall). Food samples were collected from local markets and were analyzed by ICP-MS. Cd-D thus obtained was 8.3 μg/kg body weight/month, which is equivalent to 17.1 μg/day with an assumption of 61.75 kg for an average body weight. Wang et al. [24] studied Cd-D for an ethnic group in a mountain area. Cd in food duplicate samples were measured by use of graphite furnace AAS. Mean Cd-D was 0.33 μg/kg body weight/day, which is equivalent to 20.4 μg/day when 61.75 kg was assumed for a body weight [33].

7.4 Discussion

Joint FAO/WHO Expert Committee on Food Additives (JECFA in short) adopted 25 μg/kg body weight/month as the provisional tolerable monthly intake (PTMI) [51]. Although the value for monthly intake was set taking a long biological half-life into consideration. The PTMI is per kg body weight possibly reflecting the fact that food intake may vary depending on the body size, the tolerable intake per day was estimated for practical convenience. Taking the average body weight (men and women in combination) of 59.35, 63.9, and 61.75 μ kg for Japanese, Koreans, and Chinese, respectively [7, 32, 33] (for details, see Sect. 7.2), together with an assumption that a month has 30 days, the PTMI of 25 μg/month may be equivalent to 49.5, 53.3, and 51.5 μg/day, respectively. The current level of Cd-D appears to be about 20 μg/day in Japan (Table 7.1), about 15 μg/day in Korea (Table 7.2), and about 15–20 μg/day in China (Table 7.3). Thus, none of them is in excess of the respective equivalent dose of 50–54 μg/day, staying at the level of 40, 30, and 34%, respectively.

The other point of concern is the year-dependent trend. For this purpose, the data in Tables 7.1, 7.2, and 7.3 were subjected to semilogarithmic linear regression analyses, in which the survey years were taken on the horizontal axis and the log Cd-D (μg/day) was taken on the vertical axis (Fig. 7.1 for Japan, Fig. 7.2 for Korea, and Fig. 7.3 for China). Regression line parameters are summarized in Table 7.4. The slopes were negative (i.e., less than zero, showing a trend of time-dependent decrease) for Japan and Korea. The correlation coefficients were close to 1 and statistically quite significant.

Taking advantage of significant correlation between the survey years and Cd-D (Table 7.4), Cd-D levels in near future was estimated for Japan and Korea (Table 7.5).

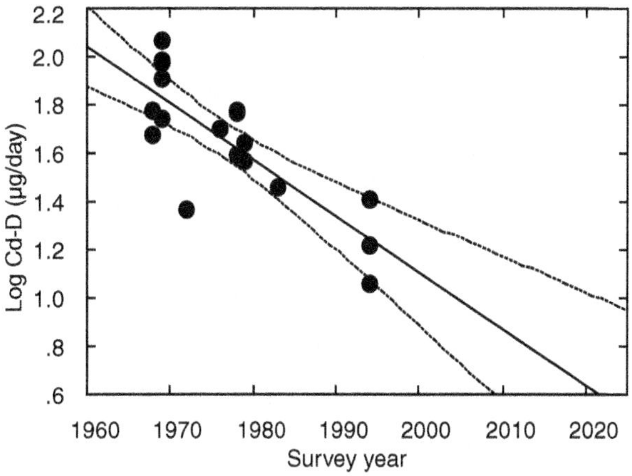

Fig. 7.1 Year-dependent trend in dietary cadmium intake (Cd-D): Japan. Survey years are on the horizontal axis, and lg Cd-D (µg/day) values are on the vertical axis. Each dot represents a GM value observed in one group of subjects. The line in the middle is a calculated regression line (for equation, see Table 7.4), and the curves on both sides of the line show the 95% ranges. For the regression line equation, see Table 7.4

Table 7.4 Regression analysis[a]

	N	Intercept (*a*)	Slope (*b*)	*r*	*p*
Japan	19	47.887	−0.023	0.809	$p < 0.001$
Korea	14	55.807	−0.027	0.899	$p < 0.001$
China	18	−30.002	0.016	0.442	$0.05 < p < 0.10$

[a]$Y = a + bX$, where X = survey year, and Y = Log Cd-D (µg/day)

Table 7.5 Estimation of Cd-D in future: Japan and Korea

	Cd-D (when *X*=)			
Method of estimation[a]	2020	2030	2040	2050
Japan				
A	26.7	15.7	9.3	5.5
B	7.5	4.9	3.1	2.0
Korea				
A	18.5	9.9	5.3	2.9
B	4.6	2.5	1.3	0.7

[a]A: Semilogarithmic correlation between *X* and *Y* where *X* = survey year and *Y* = Log Cd-D (µg/day). For details, see Table 7.4. B: Drawing of the modified exponential curve by use of SPSS. Analysis by use of Excel Tokei gave the same results

With semilogarithmic plotting (analysis A in Table 7.5), the estimation suggests that Cd-D in Korea will be 9.9 µg/day in 2030 and 5.3 µg/day in 2040. In Japan, Cd-D is estimated to be 9.3 µg/day in 2040 and 5.5 µg/day in 2050. The drawing of modified exponential curves by use of SPSS software and that by Excel Tokei software gave

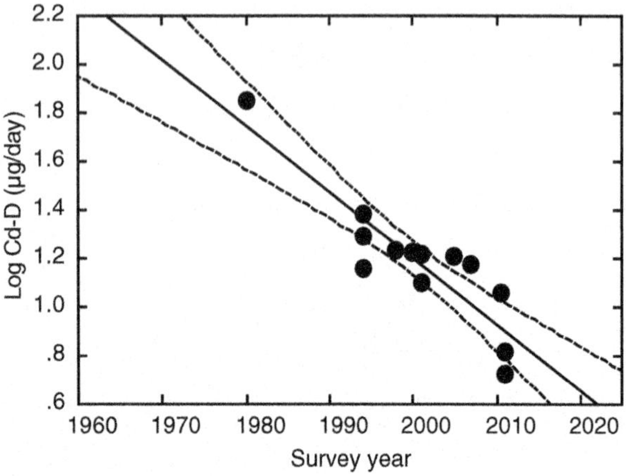

Fig. 7.2 Year-dependent trend in dietary cadmium intake (Cd-D): Korea. Legends are as under Fig. 7.1

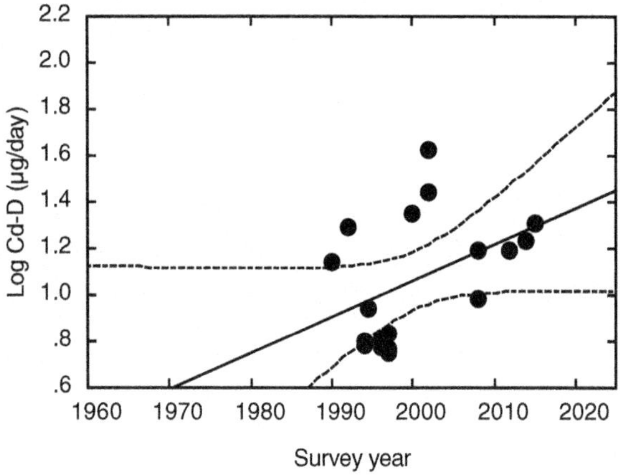

Fig. 7.3 Year-dependent trend in dietary cadmium intake (Cd-D): China. Legends are as under Fig. 7.1

essentially the same results to suggest even lower estimates for the future both in Japan and in Korea (analysis B in Table 7.5). This estimation is based on the assumption that the current trends will continue for the years to come (Figs. 7.4 and 7.5).

In 2011, the Ministry of the Environment, the Government of Japan, launched a nationwide survey to examine the effects of the environmental factors on the health of children [52]. The study involved about 100,000 pairs of newborns and their mothers in 15 sites all over the country. The target chemicals to be analyzed in bio-specimens (blood and urine from mothers) include Cd and other pollutant elements. As the project will continue for 13 or more years, Cd in bio-specimens should be available in 2024 for estimation of Cd-D for adult women in Japan.

The situation in China appears to be complex (Fig. 7.3, Table 7.3). The variation among the reported Ca-D values was wide, and statistical significance of the correlation is low. Thus, it was considered inappropriate to estimate future Cd-D. The time span from 1990 till the present may be too short, and available data may be still insufficient for the vast Chinese continent. Further studies are apparently desirable.

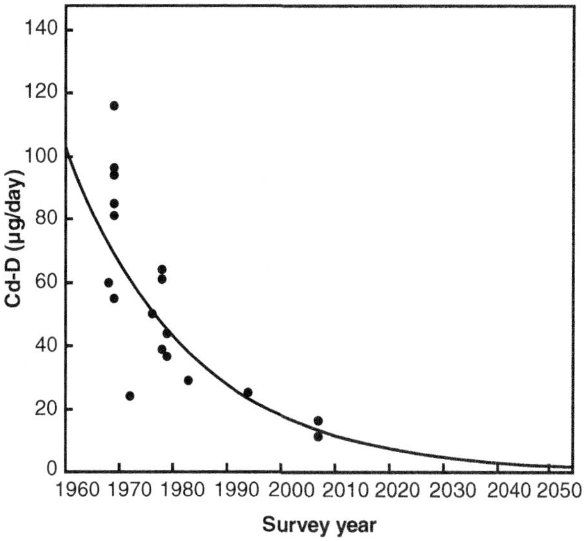

Fig. 7.4 Modified exponential curve, drawn by use of SPSS: Japan. Each dot represents a GM value observed in one group of subjects. The curve in the middle is a calculated regression curve

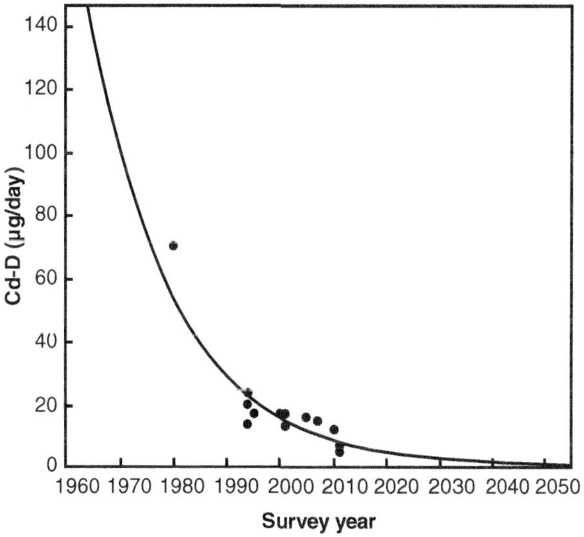

Fig. 7.5 Modified exponential curve, drawn by use of SPSS: Korea. Legends are as under Fig. 7.4

The reasons for reduction in Cd-D in Japan and Korea may be worthy of discussion. In Japan, at least three points are noted. The first is the regulation on Cd contents in marketable rice. The regulation stipulates that Cd levels in brown or polished rice should be no higher than 0.4 mg/kg (=0.4 ppm) [53, 54]. No rice with Cd higher than this limit cannot be marketed (see Chap. 8 for details of regulation in Japan).

To comply with this regulation, various improvements have been made in technology of rice cultivation in paddy. The improvements include holding of water in rice paddy till just before harvest, pH adjustment of paddy soil, and choice of rice

30. Fukui Y, Ohashi F, Sakuragi S, Moriguchi J, Ikeda M. Comparative evaluation of GFAAS and ICP-MS for analyses of cadmium in blood. Ind Health. 2011;49:338–43.
31. Sugita M, Tsuchiya K. Estimation of variation among individuals of biological half-times of cadmium calculated from accumulation data. Environ Res. 1995;68:31–7.
32. Korea Research Institute of Standard for Science. The 7th report on 'Body Size in Korea'; 2015.
33. Chinese Center for Disease Control and Prevention (eds.). The national survey on nutrition and chronic diseases China. Beijing: People's Health Press; 2015. p. 1. (in Chinese)
34. Environment Agency, the Government of Japan. Measures to treat environmental pollution with cadmium. Tokyo: The Environment Agency; 1972. p. 167–1168. (in Japanese)
35. Kitramura S. Cadmium in rice. In: Role of rice in nutrition for people in Japan; dependency on rice. Tokyo: Japan Rice Association; 1972. p. 78–126. (in Japanese).
36. Japan Public Health Association. Report on dietary intake of cadmium in cadmium-polluted areas in Japan. Tokyo: Japan Public Health Association; 1970. (in Japanese)
37. Saito H. Part 2. Survey on health of residents. In: Saito H, Takebayashi S, Harada K, Hara K, editors. Chronic cadmium poisoning. Nagasaki, Japan: Nagasaki University School of Medicine; 2003. p. 42. (in Japanese).
38. Ikeda M, Zhang Z-W, Moon C-S, Shimbo S, Watanabe T, Nakatsuka H, et al. Possible effects of environmental cadmium exposure on kidney function in the Japanese general population. Int Arch Occup Environ Health. 2000;73:15–25.
39. Ikeda M, Nakatsuka H, Watanabe T, Shimbo S. Estimation of daily cadmium intake from cadmium in blood or cadmium in urine. Environ Health Prev Med. 2015;20:455–9.
40. Moon C-S, Paik J-M, Choi C-S, Kim D-H, Ikeda M. Lead and cadmium levels in daily foods, blood and urine in children and their mothers in Korea. Int Arch Occup Environ Health. 2003;76:282–8.
41. Oh E, Lee E-I, Lim H, Jang J-Y. Human multi-route exposure assessment for Korean volunteers. J Prev Med Public Health. 2006;39:53–8. (in Korean with an English abstract)
42. Moon C-S, Yang H-R, Nakatsuka H, Ikeda M. Time trend of cadmium intake in Korea. Environ Health Prev Med. 2016;21:118–28.
43. Huang M, Choi S-J, Kim D-W, Kim N-Y, Bae H-S, Yu S-D. Evaluation of factors associated with cadmium exposure and kidney function in general population. Environ Toxicol. 2013;28:563–70.
44. Gao JQ, Li XW, Zao JL. 2000 Chinese total diet study – the dietary lead and cadmium intakes. J Hyg Res. 2006;35:750–75. (in Chinese with an English abstract)
45. Zhang Z-W, Moon C-S, Watanabe T, Shimbo S, He F-S, Wu Y-Q, et al. Back ground exposure of urban populations to lead and cadmium: comparison between China and Japan. Int Arch Occup Environ Health. 1997;69:273–81.
46. Watanabe T, Zhang Z-W, Qu J-B, Gao W-P, Jian Z-K, Shimbo S, Nakatsuka H, Matsuda-Inoguchi N, Higashikawa K, Ikeda M. Background lead and cadmium exposure of adult women in Xian city and two farming villages in Shaanxi Province. China Sci Total Environ. 2000;247:1–13.
47. Watanabe T, Zhang Z-W, Qu J-B, Xu G-F, Song L-H, Wang J-J, Shimbo S, Nakatsuka H, Higashikawa K, Ikeda M. Urban-rural comparison on cadmium exposure among general populations in Shandong Province, China. Sci Total Environ. 1998;217:1–8.
48. Jin SG. Survey report on the nutrition and health of Chinese residents: dataset on the nutrition and health in 2002. Beijing, China: People's Hygiene Press; 2008. (in Chinese)
49. Tang J, Huang Z, Pan V-D. Exposure assessment of heavy metals (cd, hg, and Pb) by the intake of local foods from Zhejiang, China. Environ Geochem Health. 2014;36:765–71.
50. He P, Lu Y, Liang Y, Chen B, Wu M, Li S, et al. Exposure assessment of dietary cadmium; findings from Shanghainese over 40 years, China. BMC Public Health. 2013;13:590. http://www.biomedcentral.com/1471-2458/13/590

51. Joint FAO/WHO Expert Committee on Food Additives (JECFA). Evaluation of the JECFA: cadmium; 2013. http://apps.who.int/food-additives-contaminants-jecfa-database/chemical.aspx?chemID=1376.
52. Kawamoto T, Nitta H, Murata K, Toda E, Tsukamoto N, Hasegawa M, et al. Rationale and study design of the Japan environment and children's study (JECS). BMC Public Health. 2014;14:25–32.
53. Ministry of Health, Labour and Welfare, the Government of Japan. Note on codes for foods, additives, etc. Tokyo: Ministry of Health, Labour and Welfare, the Government of Japan; 2010. (in Japanese)
54. Ministry of Agriculture, Forestry and Fishery, the Government of Japan. Guidance for reduction of cadmium levels in rice. Tokyo: Ministry of Agriculture, Forestry and Fishery, the Government of Japan; 2011. (in Japanese)
55. Ministry of Agriculture, Forestry and Fishery, the Government of Japan. Annual trends in rice consumption. Tokyo: Ministry of Agriculture, Forestry and Fishery, the Government of Japan; 2017. (in Japanese)
56. Matsuda-Inoguchi N, Date N, Sakurai K, Watannabe T, Toji C, Furukawa Y, et al. Limited acceptance of globalization in food habits among middle-aged women in Osaka, Japan. Int J Food Sci Nutr. 2006;57:419–33.
57. Korea Food and Drug Administration. Food code for cadmium in rice; 2012. www.foodnara.go.kr/pollution/standard/pollution01.jsp) (in Korean).
58. Korean Agency of Statistics. (http://www.index.go.kr) Consumption of main agricultural products (kg/person/year); rice; 2014. (http://www.index.go.kr/potal/main/EachDtlPageDetail.do?idx_cd=2747) (in Korean).
59. Lee S-H, Lee J-S, Choi Y-J, Kim J-G. In situ stabilization of cadmium, lead, and zinc-contaminated soil using various amendments. Chemosphere. 2009;77:1069–75.
60. Ok YS, Kim S-C, Kim D-K, Skousen JG, Lee J-S, Cheong Y-W, Kim SJ, Yang JE. Ameliorants to immobilize Cd in rice paddy soils contaminated by abandoned metal mine in Korea. Environ Geochem Health. 2011;33:23–30.

Part III
Rice Contamination by Cadmium and its Remediation

Chapter 8
Regulations for Cadmium in Rice and Soil in Japan and Countermeasures to Reduce the Concentrations

Tomoyuki Makino, Masaharu Murakami, Satoru Ishikawa, and Tadashi Abe

Abstract Rapid industrialization in the 1960s caused serious heavy metal pollution of arable soils, particularly paddy soils in some regions of Japan. The Agricultural Land Soil Pollution Prevention Law in Japan was enacted in 1970 to address heavy metal pollution. Cadmium (Cd), arsenic (As), and copper (Cu) were designated as hazardous substances targeted for regulation. Cd has been recognized as one of the most detrimental elements in Japan, due to the so-called *itai-itai* disease caused by Cd ingestion. Recently, the Codex Alimentarius Commission proposed the maximum permissible concentration of Cd in polished rice and other relevant crops. This chapter provides an overview of the regulation pertaining to Cd for soil and rice in Japan and some foreign countries. The physicochemical processes and behavior of metals in soils are described. The following conventional and advanced technologies to mitigate soil Cd contamination are also presented and evaluated: (1) water management for reducing bioavailability of soil Cd to rice plants, (2) covering and/or replacement of contaminated soil with nonpolluted soil, (3) phytoextraction of the polluted soil by rice and other potentially promising crops, and (4) chemical remediation of Cd-contaminated soil by soil washing with chemicals.

Keywords Rice · Soil · Administrative policy · Countermeasures

8.1 Regulating Cd in Soil and Rice

The Japanese government has made the following two laws to cope with soil pollution by hazardous metals and other chemical substances which are the Agricultural Land Soil Pollution Prevention Law and the Soil Contamination Countermeasures Law. These laws first proclaimed in 1970 and 2002, followed by

T. Makino (✉) · M. Murakami · S. Ishikawa · T. Abe
Institute for Agro-Environmental Sciences, National Agriculture and Food Research Organization (NARO), Tsukuba, Ibaraki, Japan
e-mail: tomoyuki.makino.d6@tohoku.ac.jp

S. Himeno, K. Aoshima (eds.), *Cadmium Toxicity*, Current Topics in Environmental Health and Preventive Medicine,
https://doi.org/10.1007/978-981-13-3630-0_8

Table 8.1 Target metals and concentration standards pertaining to soil in Japan

Metal	Soil contamination countermeasures law		Agricultural land soil prevention law
	Soil concentration standard[a] (mg kg^{-1})	Soil leachate standard[b,c] (mg L^{-1})	Concentration standard[c] (mg kg^{-1})
Cd	≤150	≤0.01	≤0.4 in rice
As	≤150	≤0.01	<15 in soil (paddy fields only)[d]
Cu	No designation	No designation	<125 in soil (paddy fields only)[e]
Cr(VI)	≤250	≤0.05	No designation
Pb	≤150	≤0.01	No designation
Hg	≤15	≤0.0005	No designation
Alkyl Hg	No designation	Not detectable	No designation
Se	≤150	≤0.01	No designation

[a]Extracted with 1 mol L^{-1} HCl, soil/solution (w/v) % = 3
[b]Extracted with water, soil/water (w/v) % = 10
[c]Analysis methods and standard values are identical to those in MOE 1991
[d]Extracted with 1 mol L^{-1} HCl, soil/solution = 10 g/50 mL
[e]Extracted with 0.1 mol L^{-1} HCl, soil/solution = 10 g/50 mL

effecting in 1971 and 2003, respectively [1, 2]. The latest revisions of the laws have been made in 2011 and 2017, respectively. Various decrees and ministerial ordinances have been issued in addition to the two laws. The maximum permissible concentrations of heavy metals regulated by the two laws at present are summarized in Table 8.1.

The designated standards for the hazardous substances in the Soil Contamination Countermeasures Law are applicable to all soils, including agricultural soils. However, the law is applied more often to legacy sites that used to be chemical and high-tech factories, where various harmful substances were once used. These sites may threaten the health and well-being of the people living in the neighborhood and pose a potential long-term health risk.

The Cd standard in the Soil Contamination Countermeasures Law, therefore, is intended to control the heavy metal pollution of soils with regard to two factors, namely, Cd concentration in soil and Cd concentration in soil leachate. The former is devised to control the significant risk of Cd intake through direct ingestion and the latter to minimize the risk of Cd intake through polluted groundwater consumption. A water extraction test is required by the leachate standard to determine the heavy metal concentrations in underground seepage [3].

The Agricultural Land Soil Pollution Prevention Law designated Cd, As, and Cu as hazardous substances to be regulated. In the case of Cd, the polluted area is designated by the Cd concentration in rice grains rather than the soil Cd concentration. This is because bioavailable Cd in soil is affected by many factors such as water management for rice cultivation, and thus, it is very difficult to determine the maximum allowable concentration by the soil Cd content alone. The concept of this regulation is quite unique; this law might be the only one of its kind globally. It could provide an incentive for farmers to decrease Cd concentration in rice by specific cultivation initiatives, such as water management.

In July 2006, the Codex Alimentarius Commission of the Joint FAO/WHO Food Standards Program proposed that the maximum permissible concentration of Cd for polished rice be limited to 0.4 mg kg^{-1} [4]. The Food Sanitation Act in Japan thus has mandated the new value of 0.4 mg kg^{-1} of Cd in brown and polished rice [5]. The corresponding standards for Cd in rice in the EU, Australia, NZ, China, and Hong Kong are 0.2, 0.1, 0.1, 0.2, and 0.1 mg kg^{-1}, respectively.

Since an estimated 34–50% of the Cd intake by Japanese nationals is derived from rice [6], it needs to develop less costly, effective, and promising technologies to tackle the issue of Cd in polluted paddy soil.

8.2 Conventional Countermeasures to Alleviate Cd Contamination in Rice and Soil in Japan

8.2.1 Water Management

Heavy metal uptake by paddy rice, particularly Cd, is considerably influenced by the water management regime due to the redox state of the paddy soil. Many studies have reported the effectiveness on the continuous submergence of the paddy fields after heading time [7–10]. Water management is a popular and cost-effective cultural practice for alleviating rice Cd contamination in Japan. Table 8.2 shows an example of the effects of water management on Cd content in rice grains [11]. Cd absorption by rice was drastically decreased by continuous submergence of the paddy fields after heading time.

The decrease of Cd bioavailability could attribute to formation of cadmium sulfide (CdS) which is sparingly soluble compounds [8, 10].

Table 8.2 Effect of water management on Cd content in rice grains [11]

		Water management after heading-time	
Soil	Cultivation	Flooding	Drained
Soil A	Pot test	Trace	1.10
Soil B	Pot test	Trace	0.68
Soil C	Field test	0.09	0.23
Soil D	Field test	0.16	0.33

Each soil in different area of Japan is contaminated with Cd

When a paddy field is flooded, the paddy soil is rapidly reduced, and consequently, its redox potential (Eh) shifts toward a reduced state (there is a sharp decrease in Eh). In such cases, the sulfate ion is reduced to sulfide [10]. The resulting sulfide ion reacts with Cd, which is precipitated out of the soil as CdS. The precipitation of CdS, in turn, lowers the Cd concentration in the soil solution, resulting in lower amounts of bioavailable Cd for the rice plants.

The formation of CdS under submerged condition of soils was theoretically estimated by using Nernst's equation and substituting the value of log K^0 into the pH–Eh relationship of H_2S (aq)/SO_4^{2-} (Fig. 8.1) [10].

Although the theoretical estimation of CdS formation has been coincident with the decrease in soil Cd availability on paddy fields, there was a lack of direct evidence on the formation of CdS due to technological limitation. Synchrotron radiation research recently developed can provide direct information on chemical species of soil Cd under flooded condition. Two Aquent soils with different S levels were used to investigate the formation of CdS. X-ray absorption fine structure (XAFS) spectroscopy and chemical extraction demonstrated the CdS proportion in the high-S soil rapidly increased to 30% at day 4 and reached 90% at day 29, whereas CdS in the low-S soil did not exceed 35% [12]. Another XAFS study showed, however, cadmium carbonate was found as the major species at most flooding periods in the alkali soils, while a small amount of cadmium sulfide was found after the soil was flooded for longer periods [13]. They also indicate the air-dried, and the 1 day-flooded alkali soils indicated that at least 50% of Cd was bound to humic acid. These results suggest CdS formation during flooded condition could be influenced by soil pH and the amount of available S. The release of carbon dioxide through rice root may effect on the formation of cadmium carbonate. It is future tasks to clarify controlling factor on the formation of hardly soluble cadmium compounds during submerged soil condition.

Flooding the soil from the tilling stage to the head formation stage during the growth of the rice would be most effective in decreasing the Cd content of the rice grains. Flooding the paddy fields as late as possible toward harvesttime is highly advisable. However, the later the flooding, the more difficult the mechanical harvesting operation. Thus, it is important to identify a compromise between lowering the bioavailable Cd content and ensuring convenient mechanical harvesting.

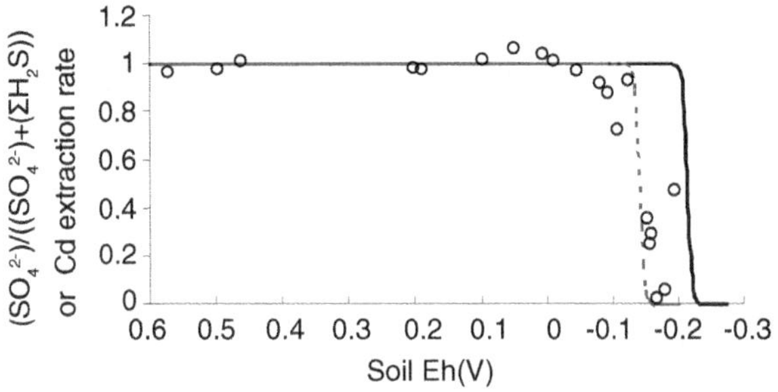

Fig. 8.1 Relationship between soil Eh and rate of Cd or sulfate residue extraction. — denotes measured values of $(SO_4^{2-})/\{(SO_4^{2-}) + (\Sigma H_2S)\}$ and the Cd extraction rate at pH 7 and ······ at pH 6 (Makino, 2002 modified from [10])

8.2.2 Soil Dressing

Soil dressing is a simple and one of the most widely used techniques for remediating heavily contaminated sites [14]. This method has been adopted as a primary countermeasure for addressing Cd contamination in agricultural soil under the Agricultural Land Soil Pollution Prevention Law in Japan. As of 2016, 92.7% of the total polluted land (7592 ha), as designated by the Agricultural Land Soil Pollution Prevention Law, was remediated by replacement with uncontaminated soil [15]. Local managers responsible for pollution prevention prefer this technique over other countermeasures because of its low risk of failure, predictable time frame, and the fact that it leaves the site in a relatively pristine condition. There are several methods to remediate polluted soils by soil dressing [16]. They include (1) covering unpolluted soil with polluted soil, (2) removing the polluted soil and refilling the space with unpolluted soil, and (3) turning the soil layers upside down (i.e., exchanging the polluted topsoil with unpolluted subsoil). According to several follow-up surveys, soil dressing is a very effective and reliable practice to decrease the Cd content in rice grains, especially when the unpolluted/newly dressed soil layer is 20–30 cm thick. However, this practice is costly and increasingly difficult to implement, recently, because of the scarcity of suitable unpolluted soils.

8.3 Countermeasures to Restore Paddy Soils Contaminated with Cd

8.3.1 Phytoremediation

Phytoremediation has drawn attention as a cost-effective and environmentally friendly technology to remove various toxic materials from soils. There are different types of phytoremediation, such as phytoextraction, phytovolatilization, phytostabilization, and rhizofiltration (Table 8.3, [17]). Phytoextraction is the most popular technology and the most intensively examined of the four phytoremediation types. A variety of plant species have been studied for their capacity to extract Cd from contaminated paddy soil, for instance, Indian mustard (*Brassica juncea*) [18, 19], sorghum (*Sorghum bicolor*) [20], hakusanhatazao (*Arabidopsis halleri* ssp. *gemmifera*) [21], sugar beet (*Beta vulgaris* L.) [18], *G. pseudochina*, *C. crepidioides*, *C. odorata*, *C. sumatrensis*, *N. tabacum* [22], maize (*Zea mays* L.), and rice (*Oryza sativa* L.) [23].

G. pseudochina, especially, highly accumulates Cd in its shoot (>71 mg Cd kg^{-1}), suggesting it could be one of the promising phytoextraction plants for paddy soils contaminated with Cd. Cultivation procedure for the plant, however, has not been developed. It would take a long time to return the converted soil (once made suitable for the upland species) to the original paddy condition.

Table 8.3 Types of phytoremediation for inorganic constituents [17]

Type of phytoremediation	Process involved	Contaminant treated
Phytostabilization	Plants control pH, soil gases, and redox conditions in soil to immobilize contaminants. Humification of some organic compounds is expected	Proven for heavy metals in mine tailing ponds
Rhizofiltration (contaminant uptake)	Compounds are taken up or biosorbed by roots (or sorbed to algae and bacteria)	Heavy metals and radionuclides
Phytoextraction (phytoaccumulation or hyperaccumulation)	Metals and organic chemicals taken up by the plant with water, or by cation pumps, sorption and other mechanisms	Nickel, zinc, lead, chromium, cadmium, selenium, other heavy metals radionuclides
Phytovolatilization	Volatile metals are taken up, changed in species, and transpired	Mercury and selenium

Interestingly, some Japonica–Indica hybrids and Indica rice varieties possess a considerably high capacity for soil Cd absorption compared to their Japonica counterparts. In this context, rice plants are considered to be one of the most promising species for Cd phytoextraction from contaminated paddy soils. Any rice varieties, regardless of their being Indica, Japonica, or their hybrids, adapt easily to Japanese paddy soils, and almost all the relevant cultural practices for rice cultivation are familiar to rice growers. Murakami et al. [23] applied a rice cultivar, Milyang 23, to paddy soil contaminated with Cd. Milyang 23 accumulated 10–15% of the total soil Cd in its shoots. These values are much higher than those reported for *Brassica juncea* (0.09%) and *Thlaspi caerulescens* (0.06%) grown on soil containing 40 mg kg^{-1} of total Cd for 6 weeks in pots [24]. The use of Milyang 23 rice is thus promising for phytoextraction of Cd from paddy soils with low Cd pollution under aerobic soil conditions [23].

Because shoot Cd uptake by rice plants is estimated as the product of the dry weight (DW) and the Cd concentration of the rice shoots, maximizing shoot Cd uptake entails applying management practices to enhance both DW and Cd uptake by the rice shoot. Healthy growth to maximize DW requires enough water supply, especially during the tillering stage (i.e., from transplanting to 30 days before panicle initiation) [25]. However, dry soil is favorable for Cd uptake by rice plants, as mentioned above. The maximization of the phytoextraction effect in paddy fields contaminated with Cd involves a trade-off between healthy plant growth and Cd concentration in rice plants. Thus, the best water management practice entails flooding the soil during tillering, followed by draining the paddy water under oxidizing conditions until harvesttime (without irrigation after drainage); this would maximize Cd accumulation by the rice shoots [26].

Another Indica rice variety, Chokoukoku, has been tested and confirmed to be one of the most promising rice cultivars [26]. Phytoextraction with Chokoukoku that was grown for 2 years without irrigation after drainage removed 883 g Cd ha^{-1}, reduced the total soil Cd content by 38%, and reduced the grain Cd content in subsequently grown Japonica food rice by 47% without decreasing yields. The

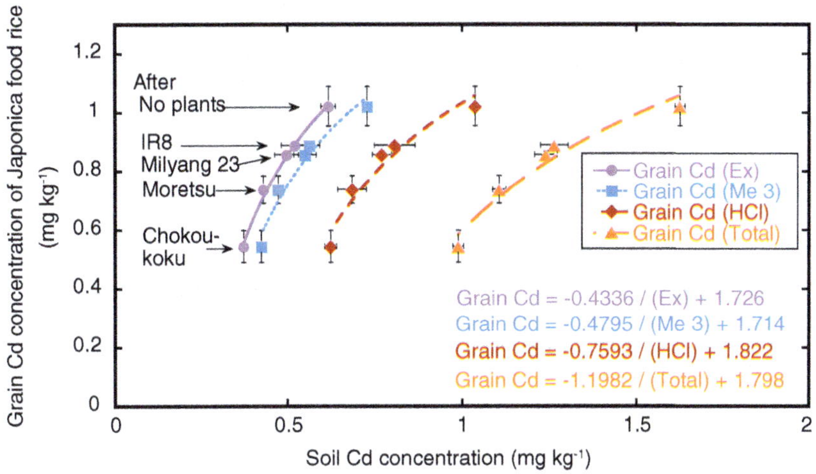

Fig. 8.2 Relationships between Cd concentrations in the exchangeable fraction (Ex), Mehlich 3-extractable fraction (Me 3), 0.1 mol L^{-1} HCl-extractable fraction (HCl), and total Cd in subplot soils of the five treatments (no plants and four indica-type rice cultivars capable of accumulating Cd at high levels), sampled before plowing in May 2007, and the corresponding grain Cd concentrations in a Japonica food rice cultivar, Yumesayaka. Soil Cd concentrations in the fractions obtained by means of Mehlich 3 extraction and 0.1 mol L^{-1} HCl extraction for each subplot

exchangeable, inorganically bound, organically bound, and total soil Cd concentrations were significantly decreased, suggesting that this cultivar can take up Cd more efficiently than the other rice cultivars from the more resistant (inorganically and organically bound) fractions, as well as from the more bioavailable (exchangeable) fraction. After the phytoextraction, the grain Cd concentrations of Japonica Yumesayaka grown after 2 years of phytoextraction (with the Indica Chokoukoku) were reduced to 47% (to 0.54 mg kg^{-1}) of that of the same rice cultivar grown without phytoextraction (1.02 mg kg^{-1}). This is the best for Cd phytoextraction result among promising rice cultivars (Fig. 8.2). These results suggest that phytoextraction with Indica Chokoukoku can remove Cd from paddy fields polluted with low to moderate levels of Cd and can reduce the grain Cd concentration of Japonica food rice cultivars to below the Codex standard within a reasonable time frame. This approach will help reduce the risk of Cd pollution of rice from paddy fields.

Although Chokoukoku showed high efficiency to remove soil Cd, the cultivar has a negative trait; it is sensitive to shattering and lodging, which are unfavorable for mechanized cropping systems in Japan. Gamma ray irradiation was applied to Jarjan seeds, an Indica rice cultivar, to produce a rice mutant that is resistant to shattering and lodging [27]. The selected mutant, named "Phyreme CD No1," has been registered under Japan's rice variety system, and a national project is ongoing to verify the effectiveness of Cd phytoextraction using this cultivar.

8.3.2 Soil Flushing

Soil washing is conventionally conducted ex situ using an appropriate apparatus with extracting reagents to remove hazardous metals from the soil into aqueous solution [28]. Due to the long distance between pollution site in agricultural land and washing treatment site in many cases, soil washing has rarely performed in pollution site in agricultural land. An on-site soil washing (soil flushing), however, could be favor for paddy fields, because an impervious hardpan is on the subsurface layer which prevents vertical percolation of water. A soil washing method practical for use on paddy field needs to meet the following criteria [29]: (1) washing chemicals should be highly efficient and generate a low environmental load, (2) the washing and wastewater treatment systems should be amenable to the paddy field's conditions and cost-effective, (3) the soil fertility of paddy fields and crop growth should not be affected considerably by the washing treatment, and (4) the effect of washing should last for a reasonably long period. We have developed a new soil washing technology that meets the abovementioned requirements.

As for washing chemicals, many researchers tried to improve extraction efficiency for soil Cd with less environmental burden, which include neutral salts, metal chelating agents, and strong acids [30]. Although neutral salts are environmentally friendly chemicals, the extraction efficiency of them is relatively low compared to other chemicals [29]. Typical metal chelating agent, ethylenediaminetetraacetic acid (EDTA), has been widely applied because of its efficacy of Cd removal from contaminated soils [31, 32]. EDTA, however, is a persistent chemical, which strongly adsorbs to soil colloids and remains in soils for a long period, providing concerns to soil environmental burden. Biodegradable chelating agents have been applied instead of EDTA because they are environmentally friendly chemicals [28, 33–36], while the cost is relatively high. It is difficult to use actual paddy fields. Although strong acidic washing solutions (HCl, HNO_3, H_2SO_4, etc.) were applied to improve the extraction efficiency for cadmium from paddy soils [37], the chemicals could provide serious soil acidification and gave fatal damage to mineralogical and physicochemical properties of paddy soils. They cannot be applied directly in paddy fields.

We, therefore, tried to select an optimal washing chemical for paddy field contaminated with Cd [29, 38]. Neutral salts, strong acids, EDTA, and iron chlorides ($FeCl_3$) were tested to evaluate their extracting efficiency of soil Cd in three paddy soils. $FeCl_3$ extracted more than 90% of the total Cd extractable by the strong acids and EDTA. Iron is one of the major soil constituents and is less environmentally harmful than either of the other chemicals, so that $FeCl_3$ was selected as a promising washing agent. Iron ions react with water to produce protons and insoluble iron hydroxide (hydrolysis reaction):

$$FeCl_3 \rightarrow Fe^{3+} + 3Cl^-$$

$$Fe^{3+} + 3H_2O \rightarrow Fe(OH)_3 + 3H^+$$

The protons released in the reactions decrease soil pH, thus enhancing the extraction efficiency of soil Cd. $FeCl_3$ can exert remarkable extraction power because of proton production through the hydrolysis. The progress of the hydrolysis reaction, unlike strong acids, is regulated by the solubility products of $Fe(OH)_3$, which provide optimal pH to extract Cd from soils without serious negative impact on soil properties.

We also conducted soil flushing in testing plots in paddy fields with $FeCl_3$ as the washing chemical. The procedure consisted of three steps: (1) chemical washing with $FeCl_3$ solution, followed by (2) water washing to eliminate the remaining chloride ion, and (3) on-site treatment of the wastewater by a portable purification apparatus with a chelating material (Fig. 8.3).

The Cd concentration in the treated wastewater was below Japan's environmental quality standard (0.003 mg Cd L^{-1}), proving that this technology is effective and promising for in situ treatment of the wastewater. The Cl concentration fell to less than 500 mg L^{-1}, the threshold value for healthy rice crops, after the soil was washed thrice with water. Determination of the chemical speciation of Cd using MINTEQ software indicated that Cd–chloride complexes were primarily formed, and this would enhance Cd extraction from the soils.

The $pH(H_2O)$ and pH(KCl) of the soil were significantly decreased by the soil flushing, while soil EC increased. They, however, did not reach the critical level at which rice growth starts slowing down. Exchangeable cation concentrations were decreased by soil washing. Total carbon and total nitrogen content were scarcely changed by washing. In summary, although changes in some relevant soil properties were observed, they can be easily corrected by the application of agricultural materials and do not affect soil fertility. The Cd contents in the washed soils were

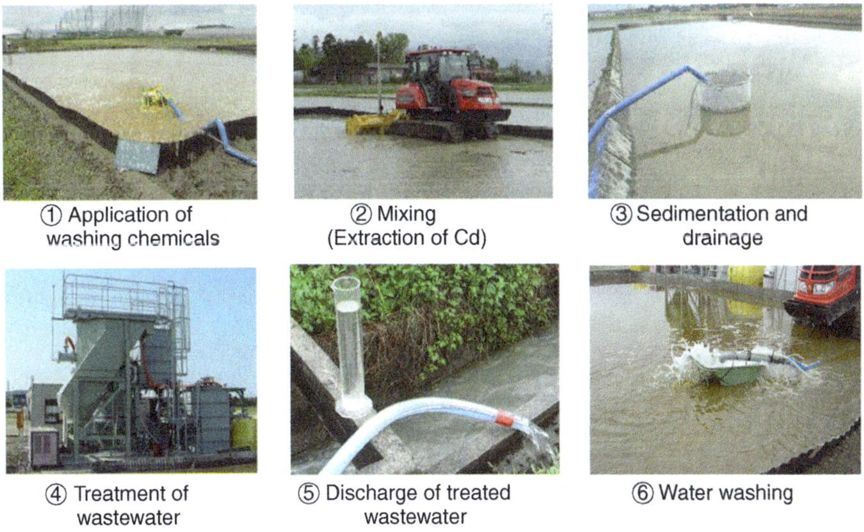

Fig. 8.3 On-site process of soil flushing with $FeCl_3$ as the washing chemical

20–40% of those in unwashed soils. The soil flushing also considerably reduced the Cd contents in the unpolished rice to 10–30% compared to those from unwashed soil. These results prove that the soil flushing technology effectively decreased the Cd contents of the polluted paddy fields.

References

1. MOE. Agricultural land soil pollution prevention law. Minist Environ Japan. 1970. https://www.env.go.jp/en/laws/water/aglaw.pdf.
2. MOE. Soil Contamination Countermeasures Act. Minist Environ Japan. 2002. https://www.env.go.jp/en/laws/water/sccact.pdf.
3. Makino T, Luo Y, Wu L, Sakurai Y, Maejima Y, Akahane I, Arao T. Heavy metal pollution of soil and risk alleviation methods based on soil chemistry. Pedologist. 2010;53:38–49.
4. Codex. Report of the 29th session of the CODEX Alimentarius Commission. Alinorm 06; 2006.
5. JETRO. Specifications and standards for foods, food additives, etc. Under the Food Sanitation Act (Abstract) 2010. Japan Extern Trade Organ. 2011:1–186.
6. Kawada T, Suzuki S. A review on the cadmium content of rice, daily cadmium intake, and accumulation in the kidneys. J Occup Health. 1998;40:264–9.
7. Arao T, Kawasaki A, Baba K, Mori S, Matsumoto S. Effects of water management on cadmium and arsenic accumulation and dimethylarsinic acid concentrations in Japanese rice. Environ Sci Technol. 2009;43:9361–7. http://www.ncbi.nlm.nih.gov/pubmed/20000530
8. Bingham FT, Page AL, Mahler RJ, Ganje TJ. Cadmium availability to rice in sludge-amended soil under 'flood' and 'nonflood' culture. Soil Sci Soc Am J. 1976;40:715–9.
9. Honma T, Ohba H, Kaneko-Kadokura A, Makino T, Nakamura K, Katou H. Optimal soil eh, pH, and water management for simultaneously minimizing arsenic and cadmium concentrations in rice grains. Environ Sci Technol. 2016;50:4178–85. https://doi.org/10.1021/acs.est.5b05424.
10. Iimura K, Ito H. Behavior and balance of contaminant heavy metals in paddy soils –studies on heavy metal pollution in paddy soils (part 2). Bull Hokuriku Natl Agri Exp Stn. 1978;21:95–145. (in Japanese)
11. Sakurai Y, Sugahara K, Makino T. Development of technology for suppressions of cadmium absorption by crops in arable soils. Kenkyuseika. 2005;434:8–14.
12. Furuya M, Hashimoto Y, Yamaguchi N. Time-course changes in speciation and solubility of cadmium in reduced and oxidized paddy soils. Soil Sci Soc Am J. 2016;80:870. https://dl.sciencesocieties.org/publications/sssaj/abstracts/80/4/870
13. Khaokaew S, Chaney RL, Landrot G, Ginder-Vogel M, Sparks DL. Speciation and release kinetics of cadmium in an alkaline paddy soil under various flooding periods and draining conditions. Environ Sci Technol. 2011;45:4249–55.
14. Vangronsveld J, Cunningham SD. Introduction to the concepts. In: Vangronsveld J, Cunningham SD, editors. Metal-contaminated soils. Berlin: Springer; 1998. p. 1–15.
15. MOE. Enforcement status of Agricultural Land-Soil Pollution Prevention Law in 2015 fiscal year. Minist Environ Japan. 2016. http://www.env.go.jp/press/files/jp/104256.pdfo_h17.pdf.
16. Yamada N. Remediation of heavy metal-contaminated soils by soil dressing, and sustainability of the remediation effects, leading edge technologies for remedying heavy metal-contaminated agricultural soils. Jpn J Soil Sci Plant Nutr. 2007;78:411–6.
17. Suthersan S. Phytoremiediateion. In: Natural and enhanced remediation systems, vol. 5. Florida: CRC Press Inc.; 2002. p. 239–67.

18. Ishikawa S, Ae N, Murakami M, Wagatsuma T. Is Brassica juncea a suitable plant for phytoremediation of cadmium in soils with moderately low cadmium contamination? - possibility of using other plant species for cd-phytoextraction. Soil Sci Plant Nutr. 2006;52:32–42.
19. Yanai J, Mabuchi N, Moritsuka N, Kosaki T. Distribution and forms of cadmium on the rhizosphere of *Brassica juncea* in cd-contaminated soils and implications for phytoremediation. Soil Sci Plant Nutr. 2004;50:423–30. http://www.tandfonline.com/doi/abs/10.1080/00380768.2004.10408496 (September, 2017)
20. Zhuang P, SHU W, LI Z, LIAO B, LI J, SHAO J. Removal of metals by sorghum plants from contaminated land. J Environ Sci. 2009;21:1432–7. https://doi.org/10.1016/S1001-0742(08)62436-5.
21. Kubota H, Sugawara R, Kitajima N, Yajima S, Tani S. Cadmium phytoremediation by Arabidopsis halleri ssp. gemmifera. Jpn J Soil Sci Plant Nutr. 2010;81:118–24.
22. Khaokaew S, Landrot G. A field-scale study of cadmium phytoremediation in a contaminated agricultural soil at Mae Sot District, Tak Province, Thailand: (1) determination of Cd-hyperaccumulating plants. Chemosphere. 2015;138:883–7. https://doi.org/10.1016/j.chemosphere.2014.09.108.
23. Murakami M, Ae N, Ishikawa S. Phytoextraction of cadmium by rice (Oryza sativa L.), soybean (Glycine max (L.) Merr.), and maize (Zea mays L.). Environ Pollut. 2007;145:96–103.
24. Ebbs SD, Lasat MM, Brady DJ, Cornish J, Gordon R, Kochian LV. Phytoextraction of cadmium and zinc from a contaminated soil. J Environ Qual. 1997;26:1424–30. https://www.agronomy.org/publications/jeq/abstracts/26/5/JEQ0260051424 (October, 2017)
25. Takahashi E. Effects of soil moisture on the uptake of silica by rice seedlings. Jpn J Soil Sci Manure. 1974;45:591–6.
26. Murakami M, Nakagawa F, Ae N, Ito M, Arao T. Phytoextraction by rice capable of accumulating cd at high levels: reduction of cd content of rice grain. Environ Sci Technol. 2009;43:5878–83. http://pubs.acs.org/doi/abs/10.1021/es8036687 (October, 2017)
27. Abe T, Kuramata M, Iwasaki Y, Honma T, Ibaraki T, Yamamoto T, Masahiro Y, Masaharu M, Satoru I. 'MJ3' and 'MA22', new high-cadmium-accumulating practical rice lines with non-shattering derived from gamma ray mutation. Breed Res. 2013;15:17–24.
28. Mulligan CN, Yong RN, Gibbs BF. Removal of heavy metals from contaminated soil and sediments using the biosurfactant surfactin. Soil Sediment Contam. 1999;8:231–54. http://www.informaworld.com/openurl?genre=article&doi=10.1080%2F10588339991339324&magic=crossref%7C%7CD404A21C5BB053405B1A640AFFD44AE3 (September, 2017)
29. Makino T, Sugahara K, Sakurai Y, Takano H, Kamiya T, Sasaki K, Itou T, Sekiya N. Remediation of cadmium contamination in paddy soils by washing with chemicals: selection of washing chemicals. Environ Pollut. 2006;144:2–10.
30. Davis A. Chemical and engineering aspects of heavy metal-contaminated soils. Rev Int Contam Ambient. 2000;16:169–74.
31. Abumaizar RJ, Smith EH. Heavy metal contaminants removal by soil washing. J Hazard Mater. 1999;70:71–86. http://linkinghub.elsevier.com/retrieve/pii/S0304389499001491 (September, 2017)
32. Zeng QR, Sauvé S, Allen HE, Hendershot WH. Recycling EDTA solutions used to remediate metal-polluted soils. Environ Pollut. 2005;133:225–31. http://www.ncbi.nlm.nih.gov/pubmed/15519453 (September, 2017)
33. Chang S, Wang K, Kuo C, Chang C, Chou C. Remediation of metal-contaminated soil by an integrated soil washing-electrolysis process. Soil Sediment Contam. 2005;14:559–69. http://www.tandfonline.com/doi/abs/10.1080/15320380500263758 (September, 2017).
34. Hong KJ, Tokunaga S, Kajiuchi T. Evaluation of remediation process with plant-derived biosurfactant for recovery of heavy metals from contaminated soils. Chemosphere. 2002;49:379–87.
35. Kantar C, Honeyman BD. Citric acid enhanced remediation of soils contaminated with uranium by soil flushing and soil washing. J Environ Eng. 2006;132:247–55. https://doi.org/10.1061/%28ASCE%290733-9372%282006%29132%3A2%28247%29.

36. Tandy S, Bossart K, Mueller R, Ritschel J, Hauser L, Schulin R, Nowack B. Extraction of heavy metals from soils using biodegradable chelating agents. Environ Sci Technol. 2004;38:937–44.
37. Ogawa B, Taguchi K, Uemura R. Studies on the mitigation of damage for crop by heavy metals in soils. Bull Akita Agri Exp Sta. 1985;35:64.
38. Makino T, Takano H, Kamiya T, Itou T, Sekiya N, Inahara M, Sakurai Y. Restoration of cadmium-contaminated paddy soils by washing with ferric chloride: cd extraction mechanism and bench-scale verification. Chemosphere. 2008;70:1035–43. http://www.ncbi.nlm.nih.gov/pubmed/17919681 (October, 2014)

Chapter 9
Molecular Mechanism of Cadmium Accumulation in Rice

Wenguang Wang, Naoki Yamaji, and Jian Feng Ma

Abstract Rice (*Oryza sativa* L.) is a major dietary source of cadmium (Cd) intake, which accounts for about 50% of total Cd intake for populations consuming rice as the staple food. Therefore, it is a very important issue to reduce transfer of Cd from soil to rice grains for human health. Recently, great progresses have been made in understanding the molecular mechanisms of Cd accumulation in rice. OsNramp5, a member of Nramp (natural resistance-associated macrophage protein) transporter family, is a major transporter for Cd uptake in rice, which is localized at the distal side of both root exodermis and endodermis. On the other hand, OsHMA3 belonging to heavy metal ATPase family is responsible for sequestration of Cd into the root vacuoles. Subsequent translocation of Cd from the roots to the shoots is mediated by OsHMA2 localized at the pericycle cells in the roots. At least two transporters (OsHMA2 and OsLCT1) localized in the nodes are involved in distribution of Cd to the rice grains. In addition, several QTL for Cd accumulation, by using wide genotypic variation in Cd accumulation, have been detected although the responsible genes remain to be identified in the future. Further identification of useful alleles for Cd accumulation will contribute to breeding rice cultivars with low Cd accumulation.

Keywords Cadmium accumulation · Distribution · Molecular mechanism · Rice grain · Transporter · Translocation · Uptake · Vacuolar sequestration

Cd is a highly toxic heavy metal for both plants and animals. Due to Cd contamination in many arable lands worldwide, its transfer from soil to the edible parts threatens our health [1–3]. This is especially a serious problem in rice because rice is a major dietary resource of Cd intake, which accounts for about 50% of total Cd intake for populations consuming rice as the staple food [4]. Furthermore, many paddy fields are mildly contaminated by Cd due to application of phosphate fertilizers and irrigation of high Cd water [2]. The "*itai-itai* disease" was caused by

W. Wang · N. Yamaji · J. F. Ma (✉)
Institute of Plant Science and Resources, Okayama University, Kurashiki, Japan
e-mail: maj@rib.okayama-u.ac.jp

S. Himeno, K. Aoshima (eds.), *Cadmium Toxicity*, Current Topics in Environmental Health and Preventive Medicine,
https://doi.org/10.1007/978-981-13-3630-0_9

consuming Cd-contaminated rice [2]. Therefore, it is very important to restrict the transfer of Cd from soil to rice grains for human health.

There are several steps for the transfer of Cd from soil to the grains such as uptake by the roots, sequestration in the roots and translocation from the roots to the shoots, and distribution to the grains. Recently, great progresses have been made in identification of transporters involved in these processes. In this chapter, the transporters involved in Cd accumulation in rice are described.

9.1 Transporters Involved in Cd Uptake in Rice Roots

Uptake of Cd by the roots from soil solution is the first step in rice Cd accumulation. Rice roots are characterized by two Casparian bands at the exodermis and endodermis and formation of aerenchyma (Fig. 9.1) [5–7]. Therefore, to cross the roots radially toward the stele for subsequent translocation of Cd from the roots to the shoots, transporters localized at the exodermis and endodermis are required (Fig. 9.1). Recently, a member of Nramp (natural resistance-associated macrophage protein) transporter family, OsNramp5, was demonstrated to be a major transporter

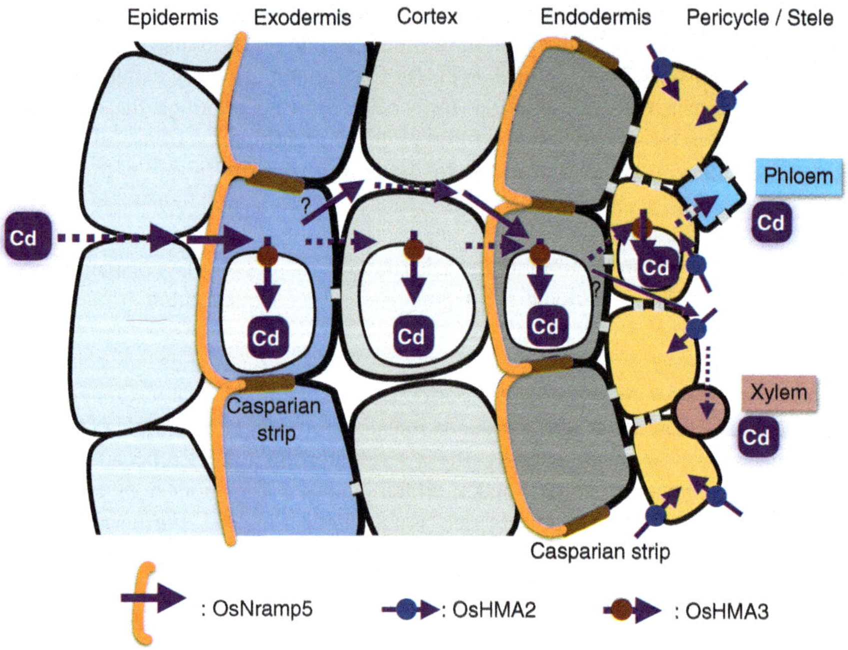

Fig. 9.1 Transporters involved in uptake and root-to-shoot translocation of Cd in rice. Cd is taken up by OsNramp5 localized at the distal side of exodermis and endodermis, followed by sequestration to the vacuoles by OsHMA3 localized at the tonoplast. OsHMA2 localized at the pericyclic cells is responsible for the root-to-shoot translocation of Cd in rice. Modified from Clemens and Ma (2016) [5]

for Cd uptake in rice roots [6, 8–10]. Nramp family is known as a transporter for divalent cations such as Zn, Mn, Fe, Cd, Ni, and Co [11–20] although one member (OsNrat1) transports trivalent Al in rice [21]. In fact, OsNramp5 is also responsible for Mn uptake in rice [6, 10]. *OsNramp5* encodes a plasma membrane-localized transporter and is mainly expressed in the rice roots through the whole growth period [6]. Its expression is not affected either by high external Mn concentrations or by essential metals such as Mn, Cu, Zn, and Fe [6]. Furthermore, *OsNramp5* is highly expressed in the mature root region. OsNramp5 protein is polarly localized at the distal side of both exodermis and endodermis cells. Knockout of *OsNramp5* resulted in complete loss of Cd uptake, indicating that OsNramp5 is a major transporter for Cd in rice. The Cd concentration in rice grains was also greatly reduced in the knockout line, but the yield was also reduced due to Mn deficiency, because Mn uptake was also hampered [6, 9]. By contrast, a different mutant (*lcd-kmt1*) of *OsNramp5* showed a similar yield with its wild type when they were grown in paddy fields although the shoot Mn concentration was also reduced in the mutant [8]. The exact reason for this inconsistency is unknown, but one possibility could be a different mutation point in the OsNramp5 protein. Nevertheless, OsNramp5 is responsible for transport of Cd from soil solution to root exodermis cells and from apoplast of aerenchyma to the endodermis cells (Fig. 9.1).

In addition to OsNramp5, OsIRT1 (iron-regulated transporter1), OsIRT2 and OsNramp1 have also been implicated in Cd uptake in rice [22, 23]. OsIRT1 and OsIRT2 belong to the ZIP (Zrt/IRT-like protein) family. They showed transport activity for Cd in yeast [22]. It was reported that excessive NO_3^- increased Cd influx in the elongation zone of rice roots by increasing *OsIRT1* expression, resulting in increased Cd uptake and accumulation in rice plants and grains [24]. On the other hand, overexpression of *OsNramp1*, a homolog of OsNramp5, resulted in increased Cd accumulation in leaves [20, 23]. However, the exact role of OsIRTs and OsNramp1 in Cd uptake remains to be further examined. Considering that knockout of *OsNramp5* completely lost Cd uptake [6, 8], the contribution of OsIRTs and OsNramp1 to the whole Cd uptake would be small.

After Cd is taken up by OsNramp5 to the root cells, Cd must be transported out of the exodermis and endodermis cells toward the xylem vessel. An efflux transporter localized at the exodermis and endodermis must be required, but it has not been identified in rice. Recently, an efflux transporter for Mn, OsMTP9, was identified [25]. It is polarly localized at the proximal side of root exodermis and endodermis, forming a cooperative uptake system with OsNramp5 for efficient Mn uptake [25]. However OsMTP9 does not transport Cd [25], indicating that a different unidentified transporter is involved in Cd efflux.

9.2 Vacuolar Sequestration of Cd in Roots

After Cd is taken up into the root cells by OsNramp5, part of Cd will be sequestered into the vacuoles. In rice roots, this is mediated by OsHMA3, a member of heavy metal-transporting ATPase family. OsHMA3 was isolated based on a large QTL for

Cd accumulation on the short arm of chromosome 7 using a mapping population derived from a cross between an *indica* variety with higher grain Cd (Anjana Dhan) and the *japonica* cultivar Nipponbare with lower grain Cd [26]. OsHMA3 is localized at the tonoplast and shows transport activity for Cd [27, 28]. *OsHMA3* is mainly expressed in the roots, and its expression is not affected by Cd exposure. OsHMA3 protein is localized at all root cells. There was no difference in the expression level, tissue, cellular, and subcellular localization of OsHMA3, between two cultivars differing in Cd accumulation. However, the OsHMA3 from high Cd-accumulating cultivar lost the transport activity for Cd due to one amino acid mutation at position 80 [27]. This mutation seems to affect the topology of OsHMA3 [27]. Therefore, in low Cd-accumulating cultivars, OsHMA3 functions to sequester Cd to the root vacuoles, resulting in less Cd translocation from the roots to the shoots and grains, while in high Cd-accumulating cultivars, loss of function of *OsHMA3* resulted in high root-to-shoot translocation of Cd (Fig. 9.1). Since the expression of native *OsHMA3* is quite low in rice roots, overexpression of functional *OsHMA3* effectively and selectively reduced Cd accumulation in the grains when grown in a Cd-contaminated soil [29]. Furthermore, there was no yield penalty in the overexpression lines. Several different loss of function alleles of *OsHMA3* were found in other rice cultivars with different genetic background [28, 30, 31]. All of these alleles showed similar phenotype, higher root-to-shoot translocation of Cd [28, 30, 31]. Therefore, these alleles may have potential for efficient phytoremediation of Cd-contaminated soil.

A recent work revealed that the original role of OsHMA3 seems a Zn transporter [29]. Overexpression of functional *OsHMA3* also increased Zn accumulation in the root cell saps including vacuoles, suggesting that OsHMA3 also functions in sequestering Zn into the vacuoles. However, different from Cd, the Zn concentration in the shoots of overexpression lines was similar to the wild type [29]. This is achieved by upregulating four *ZIP* genes involved in the Zn uptake and distribution in the *HMA3*-overexpressed lines [29]. Mutagenesis studies showed C-terminal region, particularly the first 105aa, had an important role in OsHMA3 activity [32].

9.3 Root-to-Shoot Translocation of Cd

Cd taken up by the roots will be released to the xylem for subsequent translocation to the shoots (Fig. 9.1). OsHMA2, a homolog of OsHMA3, has been reported to be responsible for the root-to-shoot translocation of Cd in rice [33, 34]. Different from OsHMA3, OsHMA2 is localized to the plasma membrane. Furthermore, OsHMA2 was localized at the pericycle of the roots. *OsHMA2* was constitutively expressed in the mature zone of the roots at the vegetative stage. OsHMA2 showed an influx transport activity for both Cd and Zn in yeast [33]. Knockout of this gene resulted in decreased Cd and Zn concentration in straw and grains [33]. However, the yield of the knockout mutants was also reduced due to decreased Zn transport. Therefore, OsHMA2 is originally responsible for the preferential distribution of Zn to the developing tissues through phloem transport.

9.4 Distribution of Cd to Rice Grains

At the reproductive growth stage, Cd will finally be distributed to the grains. Recently, node I has been demonstrated to play a crucible role in distribution of mineral elements including Cd in rice [35, 36]. Node has a complex but well-organized vascular system. There are two major vascular bundles: enlarged vascular bundles and diffuse vascular bundles [35, 36]. To deliver mineral elements to the developing tissues such as young leaves and panicles, an intervascular transfer between enlarged vascular bundles and diffuse vascular bundles is required [35, 36]. Recently, a number of transporters involved in the intervascular transfer have been identified [33, 37–42]. In rice, two transporters (OsHMA2 and OsLCT1) have been implicated in the intervascular transfer of Cd.

In addition to the roots, OsHMA2 is also expressed in the nodes (Fig. 9.2). Especially, it is highly expressed in the node I at the reproductive stage. In node I, OsHMA2 is localized at the phloem of enlarged and diffuse vascular bundles [33]. Knockout of OsHMA2 resulted in decreased concentration of zinc and cadmium in the upper nodes and reproductive organs compared with wild-type rice. Therefore, OsHMA2 is involved in reloading Zn and Cd from the intervening parenchyma tissues into the phloem of diffuse vascular bundles (Fig. 9.2).

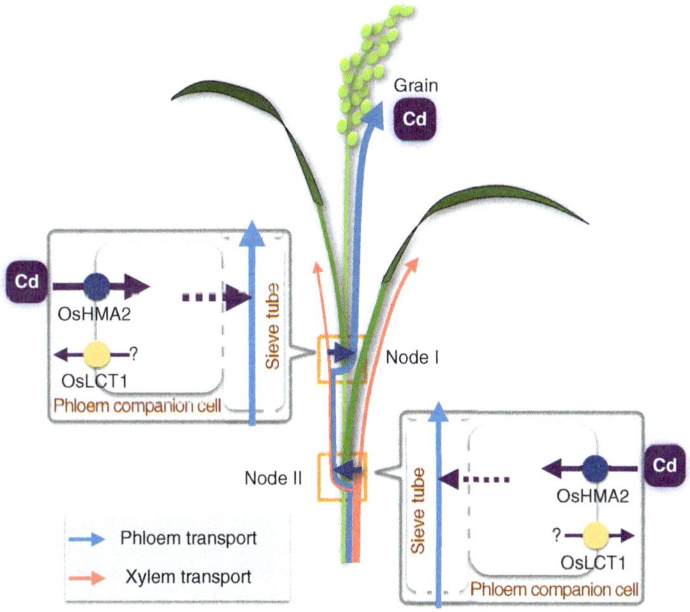

Fig. 9.2 Transporters involved in distribution of Cd in rice node. OsHMA2 localized in the phloem region of enlarged and diffuse vascular bundle is responsible for releasing Cd into the phloem. OsLCT1 is implicated in efflux of Cd from the phloem

On the other hand, OsLCT1 (*Oryza sativa* low-affinity cation transporter1) seems to be also involved in the intervascular transfer of Cd (Fig. 9.2) [37]. OsLCT1 is localized at plasma membrane and shows efflux transport activity for Cd, K, Mg, Ca, and Mn, but not for Fe, Zn, Cu, and Na [37, 43]. OsLCT1 is expressed around the EVBs and DVBs in the node I at the reproductive stage (Fig. 9.2). The expression level of *OsLCT1* was significantly higher in Kasalath than in Nipponbare [43]. Knockdown of *OsLCT1* resulted in decreased Cd concentration in phloem and grains [37]. However, in contrast to *OsHMA2* expression, which is highly expressed in the nodes throughout the reproductive growth, expression of *OsLCT1* in the nodes was only observed at the ripening stage [37]. Higher expression of *OsLCT1* was observed in the leaf blade rather than the nodes at each reproductive growth period [37]. These differences indicate that the Cd translocation pathway mediated by OsLCT1 is different from that of OsHMA2. OsLCT1 might be involved in efflux of Cd from the phloem within the nodes.

In addition to transporters described above, phytochelatin synthase 1 gene (*OsPCS1*) was also implicated in the Cd tolerance and accumulation although a contradictory result was reported [44, 45]. Knockout of *PCS1* resulted in increased sensitivity to Cd and decreased Cd accumulation in rice grain. Phytochelatin is able to form complexes with Cd, thereby detoxifying Cd internally. However, the exact mechanism for the decreased Cd accumulation remains to be investigated further.

9.5 Further Prospects

Many steps are involved in the transfer of Cd from soil to rice grains. Although several transporters involved in the uptake, translocation, and distribution of Cd have been identified in rice as described above, we are far from fully understanding of the molecular mechanism underlying the Cd accumulation. Many transporters are still missing and remain to be identified in the future by using different approaches. For example, Cd efflux transporters in the root exodermis and endodermis cells are unidentified, and transporters involved in xylem unloading and phloem loading of Cd are also unknown. Since Cd is a nonessential element for plant growth, it seems that there is no transporter specific for Cd. As described above, transport of Cd in rice is mediated by transporters for essential metals. For example, Cd uptake is mediated by OsNramp5, a Mn transporter [6, 8–10], while OsHMA2 and OsHMA3 are transporters for Zn [27, 33, 34]. Therefore, knockout of *OsNramp5* and *OsHMA2* resulted in growth inhibition due to deficiency of these essential metals. This means that these genes cannot be used for breeding rice cultivars with low Cd accumulation. Screening of the new alleles with different selectivities for Cd and essential metals may provide useful materials for breeding in the future. On the other hand, overexpression of *OsHMA3* is an efficient way to reduce Cd in rice grains [29].

There are great genotypic variations in Cd accumulation in rice [46–50]. Several QTL for Cd accumulation have been detected [26, 46, 51, 52]. Identification and pyramiding of these QTL genes in the future will contribute to breed safe rice cultivars with low Cd accumulation by marker-assisted selection. It will not be a dream to breed a rice cultivar with high Cd accumulation in the straw for phytoremediation but with low Cd in the grains for safety by manipulating transporters involved in uptake, translocation, and distribution.

Acknowledgments This work was supported by a Grant-in-Aid for Specially Promoted Research (JSPS KAKENHI grant 16H06296 to J.F.M.).

References

1. Shimo H, Ishimaru Y, An G, Yamakawa T, Nakanishi H, Nishizawa NK. *Low cadmium (LCD)*, a novel gene related to cadmium tolerance and accumulation in rice. J Exp Bot. 2011;62:5727–34. https://doi.org/10.1093/jxb/err300.
2. Horiguchi H, Aoshima K, Oguma E, Sasaki S, Miyamoto K, Hosoi Y, et al. Latest status of cadmium accumulation and its effects on kidneys, bone, and erythropoiesis in inhabitants of the formerly cadmium-polluted Jinzu River Basin in Toyama, Japan, after restoration of rice paddies. Int Arch Occup Environ Health. 2010;83:953–70. https://doi.org/10.1007/s00420-010-0510-x.
3. Zhang CH, Yin XM, Gao KH, Ge Y, Cheng WD. Non-protein thiols and glutathione S-transferase alleviate Cd stress and reduce root-to-shoot translocation of Cd in rice. J Plant Nutr Soil Sci. 2013;176:626–33. https://doi.org/10.1002/jpln.201100276.
4. Liu JG, Qu P, Zhang W, Dong Y, Li L, Wang MX. Variations among rice cultivars in subcellular distribution of Cd: the relationship between translocation and grain accumulation. Environ Exp Bot. 2014;107:25–31. https://doi.org/10.1016/j.envexpbot.2014.05.004.
5. Clemens S, Ma JF. Toxic heavy metal and metalloid accumulation in crop plants and foods. Annu Rev Plant Biol. 2016;67:489–512. https://doi.org/10.1146/annurev-arplant-043015-112301.
6. Sasaki A, Yamaji N, Yokosho K, Ma JF. Nramp5 is a major transporter responsible for manganese and cadmium uptake in rice. Plant Cell. 2012;24:2155–67. https://doi.org/10.1105/tpc.112.096925.
7. Enstone D, Peterson C, Ma FS. Root endodermis and exodermis: structure, function, and responses to the environment. J Plant Growth Regul. 2003;21:335–51. https://doi.org/10.1007/s00344-003-0002-2.
8. Ishikawa S, Ishimaru Y, Igura M, Kuramata M, Abe T, Senoura T, et al. Ion-beam irradiation, gene identification, and marker-assisted breeding in the development of low-cadmium rice. Proc Natl Acad Sci U S A. 2012;109:19166–71.
9. Yang M, Zhang Y, Zhang L, Hu J, Zhang X, Lu K, et al. OsNRAMP5 contributes to manganese translocation and distribution in rice shoots. J Exp Bot. 2014;65:4849–61. https://doi.org/10.1093/jxb/eru259.
10. Ishimaru Y, Takahashi R, Bashir K, Shimo H, Senoura T, Sugimoto K, et al. Characterizing the role of rice NRAMP5 in manganese, iron and cadmium transport. Sci Rep. 2012;2:286. https://doi.org/10.1038/srep00286.
11. Supek F, Supekova L, Nelson H, Nelson N. A yeast manganese transporter related to the macrophage protein involved in conferring resistance to mycobacteria. Proc Natl Acad Sci U S A. 1996;93:5105–10.
12. Supek F, Supekova L, Nelson H, Nelson N. Function of metal-ion homeostasis in the cell division cycle, mitochondrial protein processing, sensitivity to mycobacterial infection and brain function. J Exp Biol. 1997;200:321–30.

13. Liu XF, Supek F, Nelson N, Culotta VC. Negative control of heavy metal uptake by the *Saccharomyces cerevisiae BSD2* gene. J Biol Chem. 1997;272:11763–9.
14. Chen XZ, Peng JB, Cohen A, Nelson H, Nelson N, Hediger MA. Yeast SMF1 mediates H^+-coupled iron uptake with concomitant uncoupled cation currents. J Biol Chem. 1999;274:35089–94.
15. Nevo Y, Nelson N. The mutation F227I increases the coupling of metal ion transport in DCT1. J Biol Chem. 2004;279:53056–61. https://doi.org/10.1074/jbc.M408398200.
16. Curie C, Alonso JM, Le Jean M, Ecker JR, Briat JF. Involvement of NRAMP1 from *Arabidopsis thaliana* in iron transport. Biochem J. 2000;347:749–55.
17. Cailliatte R, Schikora A, Briat JF, Mari S, Curie C. High-affinity manganese uptake by the metal transporter NRAMP1 is essential for *Arabidopsis* growth in low manganese conditions. Plant Cell. 2010;22:904–17. https://doi.org/10.1105/tpc.109.073023.
18. Lanquar V, Lelièvre F, Bolte S, Hamès C, Alcon C, Neumann D, et al. Mobilization of vacuolar iron by AtNRAMP3 and AtNRAMP4 is essential for seed germination on low iron. EMBO J. 2005;24:4041–51.
19. Cailliatte R, Lapeyre B, Briat JF, Mari S, Curie C. The NRAMP6 metal transporter contributes to cadmium toxicity. Biochem J. 2009;422:217–28. https://doi.org/10.1042/BJ20090655.
20. Takahashi R, Ishimaru Y, Senoura T, Shimo H, Ishikawa S, Arao T, et al. The OsNRAMP1 iron transporter is involved in Cd accumulation in rice. J Exp Bot. 2011;62:4843–50. https://doi.org/10.1093/jxb/err136.
21. Xia JX, Yamaji N, Kasai T, Ma JF. Plasma membrane-localized transporter for aluminum in rice. Proc Natl Acad Sci U S A. 2010;107:18381–5.
22. Nakanishi H, Ogawa L, Ishimaru Y, Mori S, Nishizawa NK. Iron deficiency enhances cadmium uptake and translocation mediated by the Fe^{2+} transporters OsIRT1 and OsIRT2 in rice. Soil Sci Plant Nutr. 2006;52:464–9. https://doi.org/10.1111/j.1747-0765.2006.00055.x.
23. Takahashi R, Ishimaru Y, Nakanishi H, Nishizawa NK. Role of the iron transporter OsNRAMP1 in cadmium uptake and accumulation in rice. Plant Signal Behav. 2011;6:1813–6. https://doi.org/10.4161/psb.6.11.17587.
24. Yang Y, Xiong J, Chen R, Fu G, Chen T, Tao L. Excessive nitrate enhances cadmium (Cd) uptake by up-regulating the expression of *OsIRT1* in rice (*Oryza sativa*). Environ Exp Bot. 2016;122:141–9. https://doi.org/10.1016/j.envexpbot.2015.10.001.
25. Ueno D, Sasaki A, Yamaji N, Miyaji T, Fujii Y, Takemoto Y, et al. A polarly localized transporter for efficient manganese uptake in rice. Nat Plants. 2015;1:15170. https://doi.org/10.1038/NPLANTS.2015.170.
26. Ueno D, Koyama E, Kono L, Ando T, Yano M, Ma JF. Identification of a novel major quantitative trait locus controlling distribution of Cd between roots and shoots in rice. Plant Cell Physiol. 2009;50:2223–33. https://doi.org/10.1093/pcp/pcp160.
27. Ueno D, Yamaji N, Kono I, Huang CF, Ando T, Yano M, et al. Gene limiting cadmium accumulation in rice. Proc Natl Acad Sci U S A. 2010;107:16500–5.
28. Miyadate H, Adachi S, Hiraizumi A, Tezuka K, Nakazawa N, Kawamoto T, et al. OsHMA3, a P1B-type of ATPase affects root-to-shoot cadmium translocation in rice by mediating efflux into vacuoles. New Phytol. 2011;189:190–9. https://doi.org/10.1111/j.1469-8137.2010.03459.x.
29. Sasaki A, Yamaji N, Ma JF. Overexpression of *OsHMA3* enhances Cd tolerance and expression of Zn transporter genes in rice. J Exp Bot. 2014;65:6013–21. https://doi.org/10.1093/jxb/eru340.
30. Yan J, Wang P, Yang M, Lian X, Tang Z, Huang CF, et al. A loss-of-function allele of *OsHMA3* associated with high cadmium accumulation in shoots and grain of *Japonica* rice cultivars. Plant Cell Environ. 2016;39:1941–54. https://doi.org/10.1111/pce.12747.
31. Shao JF, Fujii-Kashino M, Yamaji N, Fukuoka S, Shen RF, Ma JF. Isolation and characterization of a rice line with high Cd accumulation for potential use in phytoremediation. Plant Soil. 2017;410:357–68. https://doi.org/10.1007/s11104-016-3014-y.

32. Kumagai S, Suzuki T, Tezuka K, Satoh-Nagasawa N, Takahashi H, Sakurai K, et al. Functional analysis of the C-terminal region of the vacuolar cadmium-transporting rice OsHMA3. FEBS Lett. 2014;588:789–94. https://doi.org/10.1016/j.febslet.2014.01.037.
33. Yamaji N, Xia J, Mitani-Ueno N, Yokosho K, Ma JF. Preferential delivery of zinc to developing tissues in rice is mediated by P-type heavy metal ATPase OsHMA2. Plant Physiol. 2013;162:927–39. https://doi.org/10.1104/pp.113.216564.
34. Satoh-Nagasawa N, Mori M, Nakazawa N, Kawamoto T, Nagato Y, Sakurai K, et al. Mutations in rice (*Oryza sativa*) heavy metal ATPase 2 (OsHMA2) restrict the translocation of zinc and cadmium. Plant Cell Physiol. 2012;53:213–24. https://doi.org/10.1093/pcp/pcr166.
35. Yamaji N, Ma JF. The node, a hub for mineral nutrient distribution in graminaceous plants. Trends Plant Sci. 2014;19:556–63. https://doi.org/10.1016/j.tplants.2014.05.007.
36. Yamaji N, Ma JF. Node-controlled allocation of mineral elements in Poaceae. Curr Opin Plant Biol. 2017;39:18–24. https://doi.org/10.1016/j.pbi.2017.05.002.
37. Uraguchi S, Kamiya T, Sakamoto T, Kasai K, Sato Y, Nagamura Y, et al. Low-affinity cation transporter (OsLCT1) regulates cadmium transport into rice grains. Proc Natl Acad Sci U S A. 2011;108:20959–64. https://doi.org/10.1073/pnas.1116531109.
38. Sasaki A, Yamaji N, Mitani-Ueno N, Kashino M, Ma JF. A node-localized transporter OsZIP3 is responsible for the preferential distribution of Zn to developing tissues in rice. Plant J. 2015;84:374–84. https://doi.org/10.1111/tpj.13005.
39. Yamaji N, Sasaki A, Xia JX, Yokosho K, Ma JF. A node-based switch for preferential distribution of manganese in rice. Nat Commun. 2013;4:2442. https://doi.org/10.1038/ncomms3442.
40. Zheng L, Yamaji N, Yokosho K, Ma JF. YSL16 is a phloem-localized transporter of the copper-nicotianamine complex that is responsible for copper distribution in rice. Plant Cell. 2012;24:3767–82. https://doi.org/10.1105/tpc.112.103820.
41. Deng F, Yamaji N, Xia J, Ma JF. A member of the heavy metal P-type ATPase OsHMA5 is involved in xylem loading of copper in rice. Plant Physiol. 2013;163:1353–62. https://doi.org/10.1104/pp.113.226225.
42. Yamaji N, Takemoto Y, Miyaji T, Mitani-Ueno N, Yoshida KT, Ma JF. Reducing phosphorus accumulation in rice grains with an impaired transporter in the node. Nature. 2017;541:92–5. https://doi.org/10.1038/nature20610.
43. Uraguchi S, Kamiya T, Clemens S, Fujiwara T. Characterization of OsLCT1, a cadmium transporter from *indica* rice (*Oryza sativa*). Physiol Plant. 2014;151:339–47. https://doi.org/10.1111/ppl.12189.
44. Uraguchi S, Tanaka N, Hofmann C, Abiko K, Ohkama-Ohtsu N, Weber M, et al. Phytochelatin synthase has contrasting effects on cadmium and arsenic accumulation in rice grains. Plant Cell Physiol. 2017;58:1730–42. https://doi.org/10.1093/pcp/pcx114/4082758.
45. Hayashi S, Kuramata M, Abe T, Takagi H, Ozawa K, Ishikawa S. Phytochelatin synthase OsPCS1 plays a crucial role in reducing arsenic levels in rice grains. Plant J. 2017;91:840–8. https://doi.org/10.1111/tpj.13612.
46. Ueno D, Kono I, Yokosho K, Ando T, Yano M, Ma JF. A major quantitative trait locus controlling cadmium translocation in rice (*Oryza sativa*). New Phytol. 2009;182:644–53. https://doi.org/10.1111/j.1469-8137.2009.02784.x.
47. Ishikawa S, Ae N, Sugiyama M, Murakami M, Arao T. Genotypic variation in shoot cadmium concentration in rice and soybean in soils with different levels of cadmium contamination. Soil Sci Plant Nutr. 2005;51:101–8.
48. Liu JG, Zhu QS, Zhang ZJ, Xu JK, Yang JC, Wong MH. Variations in cadmium accumulation among rice cultivars and types and the selection of cultivars for reducing cadmium in the diet. J Sci Food Agr. 2005;85:147–53. https://doi.org/10.1002/jsfa.1973.
49. Arao T, Ishikawa S. Genotypic differences in cadmium concentration and distribution of soybean and rice. JARQ. 2006;40:21–30.
50. He JY, Zhu C, Ren YF, Yan YP, Jiang D. Genotypic variation in grain cadmium concentration of lowland rice. J Plant Nutr Soil Sci. 2006;169:711–6. https://doi.org/10.1002/jpln.200525101.

51. Ishikawa S, Ae N, Yano M. Chromosomal regions with quantitative trait loci controlling cadmium concentration in brown rice (*Oryza sativa*). New Phytol. 2005;168:345–50. https://doi.org/10.1111/j.1469-8137.2005.01516.x.
52. Kashiwagi T, Shindoh K, Hirotsu N, Ishimaru K. Evidence for separate translocation pathways in determining cadmium accumulation in grain and aerial plant parts in rice. BMC Plant Biol. 2009;9:8. https://doi.org/10.1186/1471-2229-9-8.

Chapter 10
Mitigation Strategies for Cadmium and Arsenic in Rice

Tomohito Arao

Abstract Rice (*Oryza sativa* L.) is a major dietary source of not only cadmium (Cd) but also inorganic arsenic (iAs) in populations that consume rice as a staple food. When rice is cultivated under aerobic conditions, Cd in the paddy soil is solubilized and more likely to accumulate in the grain. Under flooded conditions iAs is more likely to accumulate in the grain. The application of intermittent irrigation consisting of 3 days of flooding and several days of drainage, during the heading stage of rice growth, was effective for simultaneously decreasing the accumulation of Cd and iAs in grain without negative effects on yield and quality. An alternative and promising strategy is a combination of water management and soil amendment to reduce the absorption of Cd and iAs. Growing a low-Cd-accumulating cultivar under aerobic conditions is the most practical way to simultaneously reduce the Cd and iAs contents in rice. Bacteria associated with the rice rhizosphere are involved in the formation of dimethylarsinic acid (DMA) in rice.

Keywords Arsenic · Cadmium · Inorganic arsenic · Arsenite · Rice · Water management · Soil amendment · Low-Cd cultivar · Dimethylarsinic acid

10.1 General

Arsenic (As) is a metalloid with diverse organic and inorganic forms that are widely distributed in water, soil, and air. Small amounts of organic As and inorganic As (iAs) are contained in most foodstuffs. According to the European Commission Scientific Cooperation Project [1], intake of As from marine foodstuffs may be in excess of 50%. As is present mostly as arsenobetaine in fish and shellfish [2] and mostly as arsenosugar in seaweed [3], both of these species are considered virtually nontoxic. iAs species (arsenate, As(V), and arsenite, As(III)) have greater toxicity than organic As species, and iAs is a nonthreshold class 1 carcinogen in humans based on epidemiologic studies in groundwater-contaminated regions [4].

T. Arao (✉)
Central Region Agricultural Research Center, NARO, Tsukuba, Ibaraki, Japan
e-mail: arao@affrc.go.jp

S. Himeno, K. Aoshima (eds.), *Cadmium Toxicity*, Current Topics in Environmental Health and Preventive Medicine,
https://doi.org/10.1007/978-981-13-3630-0_10

While there is concern about the health risks associated with high levels of iAs in drinking water in many parts of the world, iAs in rice is the largest dietary source for the portion of the world's population without elevated iAs levels in their drinking water [5, 6]. Rice contains elevated As concentrations compared to wheat and barley [7]. A market-basket study indicated that in Japan, cereals contributed the most to the total daily intake of iAs, followed by algae [8]. Rice and rice cakes contributed 97% of the iAs intake from cereals, and virtually 100% of iAs in algae was from hijiki (*Sargassum fusiforme*). Thus, two food items—rice and hijiki—are major contributors to the daily iAs intake in the Japanese population. Using a probabilistic approach, the daily bioaccessible iAs intake in Japan was estimated to be 19 and 59 μg/day at 50 and 95 percentiles, respectively [9].

Rice is a major dietary source of not only iAs but also Cd in populations that consume rice as a staple food. Approximately 40% of the Cd intake of the Japanese is derived from rice [10]. The Codex Alimentarius Commission set the maximum allowable levels of Cd and iAs in polished rice as 0.4 mg kg^{-1} in 2008 [11] and 0.2 mg kg^{-1} in 2016 [12], respectively. The Commission also set the maximum allowable level for iAs in brown rice at 0.35 mg kg^{-1}. According to a survey of rice As concentrations conducted by the Ministry of Agriculture, Forestry and Fisheries of Japan (MAFF), 2.2% of polished rice cultivated in Japan in 2012 exceeded 0.20 mg kg^{-1} iAs, and the average value ($n = 600$) of iAs was 0.12 mg kg^{-1} in polished rice and 0.21 mg kg^{-1} in brown rice; the latter of which represents an increase compared to the level of 0.15 mg kg^{-1} in brown rice detected in 2004–2006 [13].

When rice plants are cultivated under aerobic conditions, Cd in paddy soil is more easily solubilized and accumulated in rice compared to the solubilization and accumulation under anaerobic conditions. In contrast, more iAs accumulates in rice under flooded conditions [14] because paddy cultivation mobilizes iAs in the soil through the reductive dissolution of iron (Fe) oxide and the subsequent reduction of As(V) to the more bioavailable As(III) [15].

The effects of polishing, cooking, and storage on the total As and As species concentrations in rice were studied under typical conditions in Japan [16]. The total As and iAs levels in three white rice samples polished by removing 10% of the outer layers from brown rice were reduced to 61–66% and 51–70%, respectively, of the levels in brown rice. The total As and iAs levels in polished rice after three washings with deionized water were reduced to 81–84% and 71–83%, respectively, of the levels in raw rice. Low-volume cooking (water/rice of 1.4–2.0:1) of rice to dryness did not remove As. The As content of brown rice stored in grain form for 1 year was stable.

10.2 Water Management

As and Cd solubilization/immobilization in paddy soils is influenced by reduction and oxidation processes that occur in response to alternate submergence and drainage practices in the field. Both the As solubilization under reducing conditions and immobilization under oxidizing conditions are coupled with the concurrent Fe reduction/oxidation processes. The relationships between the dissolved As and Cd concentrations and air-filled porosity suggest that there is an optimum soil wetness that simultaneously minimizes the solubilization of both As and Cd [17].

The factors controlling the soil-to-solution partitioning of As under anaerobic conditions were investigated. Changes in As and Fe speciation due to the flooded incubation of two paddy soils (soils A and B) were investigated by high-performance liquid chromatography-inductively coupled plasma mass spectrometry (HPLC-ICPMS) and X-ray absorption near-edge structure analysis [18]. The flooded incubation decreased the Eh, increased the pH, and increased the As(III) fraction in the solid soil phase to 80% of the total As in the soils. The solution-to-soil ratios of As(III) and As(V) indicated that As(III) was more readily released from soil to solution than was As(V). The amount of As dissolved in anaerobic incubation was lower in soil A than in soil B. The dissolution of Fe minerals rather than the redox conversion of As(V) to As(III) explained the different dissolution amounts of As in the two paddy soils. The ratio of As to Fe released from the soils decreased with the incubation time during both anaerobic incubation and abiotic dissolution by sodium ascorbate, suggesting that a larger amount of As was associated with an easily soluble fraction of Fe hydroxide in the amorphous phase and/or smaller particles.

The effects of water management in rice paddies on the levels of Cd and As in rice grain were investigated in pot experiments [14]. Flooding increased the As concentrations in rice grain, whereas aerobic treatment increased the concentration of Cd. Flooding for 3 weeks before and after heading was most effective at reducing the grain Cd concentration, but this treatment increased the As concentration considerably. Conversely, aerobic treatment during the same period was effective at reducing the As concentrations but increased the Cd concentration markedly. The concentration of dimethylarsinic acid (DMA) in grain was very low under aerobic conditions but increased under flooded conditions. In the concrete frame experiments, the grain As and Cd concentrations were higher and lower, respectively, under flooding for 3 weeks before and after heading than under intermittent irrigation [19].

The effect of different start periods of drainage after heading on the levels of Cd and As in rice grain was examined in concrete frames filled with soil collected from an area surrounding a former As-polluted region [20]. The concentration of As and Cd in grain decreased and increased, respectively, with an early start to the drainage. A field experiment was conducted to investigate the effects of four water management regimes on As and Cd accumulation [21]. Water management both before and after the full tillering stage affected As and Cd accumulation in the grains.

Figure 10.1a illustrates water management in typical paddy fields in Japan. The paddy fields are submerged for 1 month after transplanting, followed by midseason drainage for approximately 20 days. After the midseason drainage, intermittent irrigation is usually applied for 2 months, with an exception of 1 week of submergence at the heading stage. The fields are drained 2 weeks before the harvest.

Field experiments were conducted to investigate the optimal conditions of Eh and pH in soil that simultaneously decrease iAs and Cd accumulation in rice [22]. Water

with WTR affected dissolved As in flooded soils and As uptake by rice plants [24]. Compared to the As concentrations in rice plants grown on control soil, the As concentrations in plants grown on WTR-treated soils decreased by approximately 20% in the grain.

Fe plaque, the coating of Fe oxide on rice roots, may affect As uptake. Fe plaque is generally considered to be a buffer or barrier to As uptake because of the plaque's high affinity for As. When rice was grown under anaerobic conditions, Fe plaque coatings were not observed on young rice roots or on the younger portions of mature roots, which are the most active areas of solute uptake, whereas distinct pigmentation by Fe plaque was found on mature roots [32]. When the soil matrix was anaerobic during rice growth, Fe plaque did not cover the entire root, and As(III) was the dominant As species in the soil matrix and rhizosphere [33]. Draining before harvest shifted the conditions to aerobic. The oxidation of As(III) to As(V) occurred faster in the Fe plaque than in the soil matrix. As was scavenged by Fe mottles originating from the Fe plaque around the roots. The ratio of As(V) to As(III) decreased toward the outer rim of the subsurface Fe mottles, where the soil matrix was not completely aerated. These results provide direct evidence that the speciation of As near rice roots depends on spatial and temporal redox variations in the soil matrix.

Increasing plant available silica (Si) can decrease grain As because As(III) shares a root uptake transporter with Si [34]. The incorporation of Si-rich rice husk into soil (1% w/w, a 10-fold higher husk amount than would be available each crop cycle) decreased grain iAs by 25–50% without negatively affecting grain Cd in pot experiments [35]. However, following the application of calcium silicate slag at a rate of 0.5 kg m^{-2} in concrete frame experiments, no significant effect was observed on the As concentration in rice grain or straw, but the available silicate in the soil increased [29]. It was suggested that a higher application rate of calcium silicate slag than that in the study would be needed to mitigate As accumulation in rice grain. Further research is needed on Si amendment to decrease As accumulation in rice grain.

The effects of three types of organic amendments on As solubility during long-term flooded soil incubation were comprehensively examined [36]. The amendments examined were categorized as (1) plant-based composts, (2) cattle manure, and (3) fermented livestock droppings. The dissolved As in all soils with fermented livestock dropping was significantly greater than that in control soils at 10 and 39 days after incubation (Fig. 10.2). Furthermore, the solubilizing effect of some organic amendments was mostly explained by promotion of the decrease in soil Eh, rather than the increase in dissolved organic carbon. Thus, fermented livestock droppings potentially have a greater risk of solubilizing As than plant-based composts and cattle manure, because of the greater ability of the droppings to accelerate soil reduction, which is associated with their higher bio-decomposability.

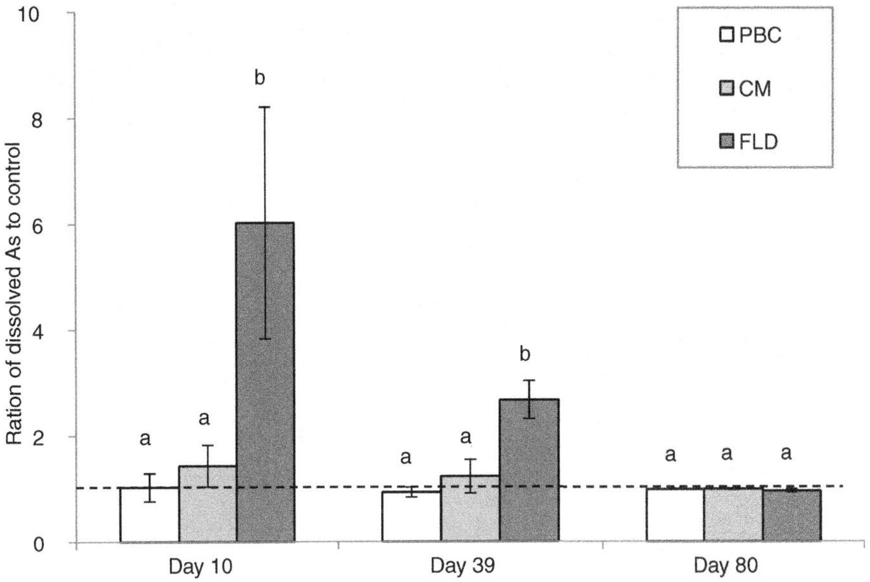

Fig. 10.2 Average relative ratios of dissolved As in soils with each type of organic amendments and in control soils [36]. The broken line indicates a ratio of 1. PBC, CM, and FLD indicate plant-based compost, cattle manure, and fermented livestock dropping, respectively. Error bars indicate the mean ± standard deviation. Different letters on the same day denote significant differences at $P = 0.05$ (Tukey's test)

10.4 Low Cd Variety

A japonica rice mutant with nearly undetectable levels of Cd in the grain was produced [37]. The name "Koshihikari Kan No. 1" was chosen, and in 2015 the cultivar was registered in Japan. This cultivar and "Koshihikari," its parent and the most popular cultivar in Japan, were grown in paddy fields with different soil properties under three water regimes: flooded conditions (FLD), alternate wetting and drying conditions (AWD), and water-saving conditions (WAS). The application of FLD for several weeks before and after heading significantly increased the grain As concentration in both cultivars. AWD, with the soil re-flooded just after the disappearance of ponded water, reduced the grain As concentrations by an average of 27% relative to those under FLD for both cultivars. WAS, with irrigation after drying of the soil surface, decreased the grain As concentrations by an average of 43–48% compared to FLD. Although AWD and WAS remarkably increased grain Cd concentrations in "Koshihikari," "Koshihikari Kan No. 1" had nearly undetectable levels of grain Cd in all treatments (Fig. 10.3, [38]). Although aerobic conditions such as WAS have slightly adverse effects on grain yield and quality, growing the low-Cd cultivar aerobically is the most practical way to simultaneously reduce Cd and As contents in the rice grain.

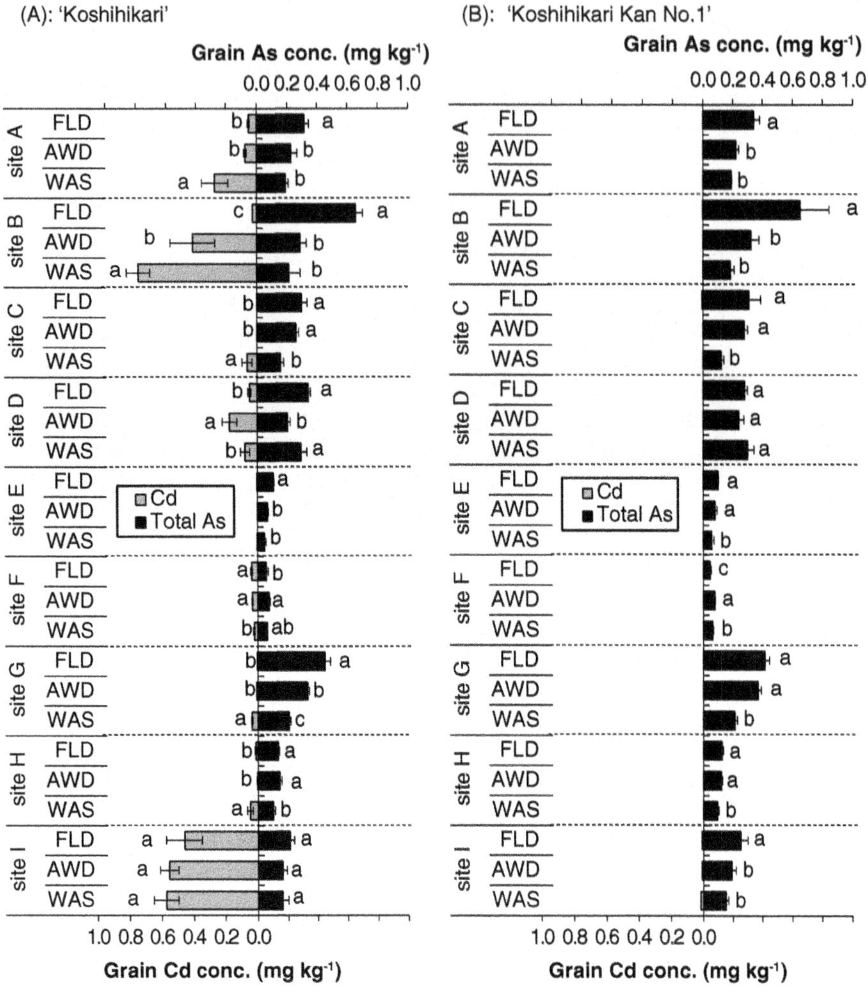

Fig. 10.3 Effects of the three water regimes on the concentrations of As and Cd in the grains of "Koshihikari" and "Koshihikari Kan No. 1" grown at the nine experimental sites in Japan [38]. *FLD* flooded conditions, *AWD* alternate wetting and drying conditions, *WAS* water-saving conditions. Values are the means ± standard deviation (SD; $n = 3$–4). Bars at a site labeled with different letters differ significantly (Tukey's honestly significant difference test, $p < 0.05$)

DNA markers that can be used to introduce the gene responsible for low Cd uptake into various cultivars have been developed [37], and breeding programs have been launched to transfer the low-Cd trait into other popular cultivars in Japan.

10.5 Low As Variety

As accumulation and speciation in the major cultivars currently grown in Japan were examined [39]. Ten major rice cultivars were grown under flooded conditions in a paddy field with a background level of As (low-As soil) or in pots filled with soil containing a high level of As (high-As soil). In the low-As soil, there were few genotypic differences in the levels of either total As or iAs in the grain. In the high-As soil, the total grain As increased to a mean level of 2.4 mg kg^{-1} in the ten cultivars, with markedly increased levels of DMA. The genotypic variability of iAs levels was quite small, and these levels remained low (at approximately 0.2 mg kg^{-1}) even when the total As levels increased markedly.

The genetic diversity in As accumulation and As speciation in rice grain was investigated using a world rice core collection (WRC) comprising 69 accessions grown over a 3-year period (Fig. 10.4). There was a threefold difference in the As concentration in the grains from the WRC. The concentrations of total As, iAs, and DMA were significantly affected by the genotype, the year, and the genotype-year interaction effects. Among the WRC accessions, Local Basmati and Tima (*indica* type) were identified as cultivars with the lowest stable total As and iAs concentrations. Using an F2 population derived from Padi Perak (a high-DMA cultivar) and Koshihikari (a low-DMA cultivar), two QTLs on chromosome 6 and one QTL on chromosome 8 that were responsible for variations in the grain DMA concentration were identified.

Mutation in Lsi1 (a Si influx transporter) significantly decreases As(III) uptake [34]. In the rice mutants defective in the silicon efflux transporter Lsi2, As(III) transport to the xylem and accumulation in shoots and grain decreased greatly. Mutation in Lsi2 had a much greater impact on As accumulation in the shoots and grain of field-grown rice than Lsi1. However, mutation in Lsi1 or Lsi2 would decrease Si uptake and could have a deleterious influence on growth.

OsABCC1 is localized in the tonoplast of phloem cells in the node and is responsible for the sequestration of As(III)–phytochelatin into the vacuoles [40]. When this transporter was knocked out, less As accumulated in the node, and more As was distributed to the grain. Overexpression of OsABCC1 may be used as a strategy to breed low As-accumulating rice cultivars. The inhibition of thiol synthesis by L-buthionine sulfoximine markedly decreased As accumulation in the node but increased As distribution to the flag leaf and the grain [41]. Transgenic plants overexpressing the phytochelatin synthase OsPCS1 showed significantly lower As levels in the grain than did wild-type plants [42].

In *Saccharomyces cerevisiae*, ScAcr3p, an As(III) plasma membrane transporter, mediates the efflux of As(III) and confers tolerance to arsenicals. The expression of ScACR3 in rice enhanced As(III) efflux activity by 30% and thus reduced As accumulation in brown rice by 20% [43]. Mechanistic insights into the accumulation of As and Cd in crop plants are reviewed by Clemens and Ma [44].

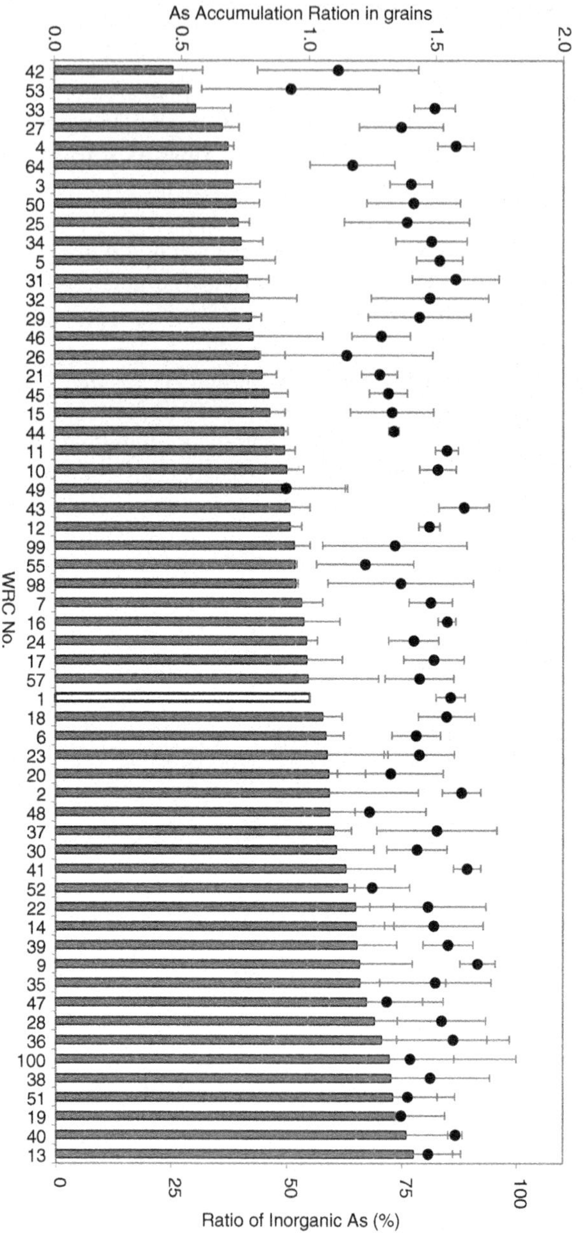

Fig. 10.4 Genotypic variation in the As accumulation in grains from the world rice core collection (WRC) comprising 69 accessions [50]. Grains from the WRC were cultured for over 3 years in a paddy field containing naturally abundant As. The accessions that did not ripen because of late heading were omitted from the As analysis, and 58 accessions are presented. The columns reflect the ratio of the total As concentration (filled columns) in each genotype to that in WRC1 (open column), japonica cv. Nipponbare, averaged over 3 years. Circles indicate the average proportions of iAs in the sum of As species in each genotype

10.6 Dimethylarsinic Acid (DMA)

Most As is present in rice grain as iAs (mainly As(III)) and DMA. The grain of aerobically grown rice (cv. "Koshihikari") contains As almost exclusively in the form of iAs, whereas in anaerobically grown rice, DMA accounts for the majority of the total As [14]. Whether DMA in rice is taken up from the soil or synthesized from iAs in plants remains unresolved. The effect of DMA, MMA, and As(III) amendment on As speciation in rice grain grown in soil and in solution culture was investigated [45]. When rice was grown under flooded conditions after the heading stage, DMA amendment to the soil resulted in higher DMA concentration in brown rice and rice straw. In the solution culture, not only DMA amendment but also MMA or As(III) amendment increased the DMA concentration in brown rice and rice straw. DMA was detected in the solution containing young rice plants amended by MMA or As(III). When the solution included the antibacterial agent chloramphenicol, the DMA concentration in the solution decreased dramatically. When only the soil was incubated with MMA or arsenite, only a small amount of DMA was detected in the soil. These results suggest that bacteria associated with the rice rhizosphere are involved in the formation of DMA in brown rice.

The existence of processes that transport DMA from the soil to rice grain is supported by two pieces of evidence: the recognition of methylated As by silicon transporters [46] and the translocation of methylated As to the rice grain via the xylem and phloem with an efficiency at least an order of magnitude greater than that for iAs [47]. Methylated arsenicals were not found in the three plant species exposed to iAs under axenic conditions. Axenically grown rice was able to take up MMA but not convert it to DMA [48].

Isolation and functional analysis of the microbes mediating the methylation of As in paddy soils is important for understanding the origin of DMA in rice grain. Strain GSRB54 responsible for As methylation was isolated from the rice rhizosphere [49]. DMA was detected in the shoots of rice grown in liquid medium inoculated with GSRB54 and containing As(III). Since *Streptomyces* are generally aerobic bacteria, strain GSRB54 likely inhabits the oxidative zone around the roots of paddy rice and is associated with DMA accumulation in rice grain through As methylation in the rice rhizosphere.

References

1. SCOOP (Scientific Co-operation on Questions Relating to Food). Assessment of dietary exposure to arsenic, cadmium, lead, mercury of the population of the European Union Member States; 2004. https://ec.europa.eu/food/sites/food/files/safety/docs/cs_contaminants_catalogue_scoop_3-2-11_heavy_metals_report_en.pdf.
2. Hanaoka K, Yamamoto H, Kawashima K, Tagawa S, Kaise T. Ubiquity of arsenobetaine in marine animals and degradation of arsenobetaine by sedimentary microorganisms. Appl Organometal Chem. 1988;2:371–6.

3. Francesconi KA, Edmonds JS. Arsenic and marine organisms. Adv Inorg Chem. 1997;44:147–89.
4. IARC (International Agency for Research on Cancer). A review of human carcinogens-Part C: Arsenic, Metals, Fibres and Dusts. In: IARC monographs on the evaluation of carcinogenic risks to humans, vol. 100C. Lyon, France: International Agency for Research on Cancer; 2012. p. 41–93.
5. EFSA. Scientific opinion on arsenic in food. EFSA J. 2009;7:1351.
6. Meharg AA, Zhao FJ. Arsenic & rice. London: Springer; 2012.
7. Williams PN, Villada A, Deacon C, Raab A, Figuerola J, Green AJ, Feldmann J, Meharg AA. Greatly enhanced arsenic shoot assimilation in rice leads to elevated grain levels compared to wheat and barley. Environ Sci Technol. 2007;41:6854–9.
8. Oguri T, Yoshinaga J, Tao H, Nakazato T. Inorganic arsenic in the Japanese diet: daily intake and source. Arch Environ Contam Toxicol. 2014;66:100–12.
9. Oguri T, Yoshinaga J. Daily inorganic arsenic intake of the japanese estimated by a probabilistic approach. Jpn J Hyg. 2014;69:177–86. (in Japanese)
10. MAFF. Survey of cadmium levels in agricultural products in Japan and estimation of cadmium intake in Japan. 2016. http://www.maff.go.jp/j/press/syouan/nouan/pdf/160223-01.pdf. Accessed 30 Jun 2017, (in Japanese).
11. Codex Alimentarius Commission In. Codex stan 193-1995, Codex general standard for contaminants and toxins in foods and feed. 2008. http://www.fao.org/fileadmin/user_upload/livestockgov/documents/1_CXS_193e.pdf. Accessed 30 Jun 2017.
12. Codex Alimentarius Commission In. REP16/CF Joint FAO/WHO Food Standards Programme Codex Alimentarius Commission 39th Session Rome, Italy, 27 June–1 July 2016. http://www.fao.org/fao-who-codexalimentarius/sh-proxy/zh/?lnk=1&url=https%253A%252F%252Fworkspace.fao.org%252Fsites%252Fcodex%252FMeetings%252FCX-735-10%252FReport%252FREP16_CFe.pdf. Accessed 30 Jun 2017.
13. MAFF. Survey of arsenic levels in brown rice and polished rice produced in Japan. 2014. http://www.maff.go.jp/j/press/syouan/nouan/pdf/140221-01.pdf. Accessed 30 Jun 2017, (in Japanese).
14. Arao T, Kawasaki A, Baba K, Mori S, Matsumoto S. Effects of water management on cadmium and arsenic accumulation and dimethylarsinic acid concentrations in Japanese rice. Environ Sci Technol. 2009;43:9361–7.
15. Takahashi Y, Minamikawa R, Hattori KH, Kurishima K, Kihou N, Yuita K. Arsenic behavior in paddy fields during the cycle of flooded and non-flooded periods. Environ Sci Technol. 2004;38:1038–44.
16. Naito S, Matsumoto E, Shindoh K, Nishimura T. Effects of polishing, cooking, and storing on total arsenic and arsenic species concentrations in rice cultivated in Japan. Food Chem. 2015;168:294–301.
17. Nakamura K, Katou H, Suzuki K, Honma T. Air-filled porosity as a key to reducing dissolved arsenic and cadmium concentrations in paddy soils. J Environ Qual. 2018;47:496–503.
18. Yamaguchi N, Nakamura T, Dong D, Takahashi Y, Amachi S, Makino T. Arsenic release from flooded paddy soils is influenced by speciation, Eh, pH, and iron dissolution. Chemosphere. 2011;83:925–32.
19. Kawasaki A, Arao T, Ishikawa S. Reducing cadmium content of rice grains by means of flooding and a few problems. Jpn J Hyg. 2012;67:478–83. (in Japanese)
20. Matsumoto S, Kasuga J, Taiki N, Makino T, Arao T. Reduction of the risk of arsenic accumulation in rice by the water management and material application in relation to phosphate status. J Plant Inter. 2015a;10:65–74.
21. Hu P, Huang J, Ouyang Y, Wu L, Song J, Wang S, Li Z, Han C, Zhou L, Huang Y, Luo Y, Christie P. Water management affects arsenic and cadmium accumulation in different rice cultivars. Environ Geochem Health. 2013;35:767–78.

22. Honma T, Ohba H, Kaneko-Kadokura A, Makino T, Nakamura K, Katou H. Optimal soil Eh, pH, and water management for simultaneously minimizing arsenic and cadmium concentrations in rice grains. Environ Sci Technol. 2016a;50:4178–5.
23. Suda A, Baba K, Yamaguchi N, Akahane I, Makino T. The effects of soil amendments on arsenic concentrations in soil solutions after long-term flooded incubation. Soil Sci Plant Nutri. 2015;61:592–602.
24. Suda A, Baba K, Akahane I, Makino T. Use of water treatment residue containing polysilicate iron to stabilize arsenic in flooded soils and attenuate arsenic uptake by rice *Oryza sativa*. Soil Sci Plant Nutri. 2016;62:111–6.
25. Komárek M, Vaněk A, Ettler V. Chemical stabilization of metals and arsenic in contaminated soils using oxides - a review. Environ Pollut. 2013;172:9–22.
26. Bang S, Johnson MD, Korfiatis GP, Meng X. Chemical reactions between arsenic and zero-valent iron in water. Water Res. 2005;39:763–70.
27. Sun H, Wang L, Zhang R, Sui J, Xu G. Treatment of groundwater polluted by arsenic compounds by zero valent iron. J Hazard Mater. 2006;129:297–303.
28. Kumpiene J, Ore S, Renella G, Mench M, Lagerkvist A, Maurice C. Assessment of zerovalent iron for stabilization of chromium, copper, and arsenic in soil. Environ Pollut. 2006;144:62–9.
29. Matsumoto S, Kasuga J, Taiki N, Makino T, Arao T. Inhibition of arsenic accumulation in Japanese rice by the application of iron and silicate materials. Catena. 2015b;135:328–35.
30. Matsumoto S, Kasuga J, Makino T, Arao T. Evaluation of the effects of application of iron materials on the accumulation and speciation of arsenic in rice grain grown on uncontaminated soil with relatively high levels of arsenic. Environ Exp Bot. 2016;125:42–51.
31. Makino T, Nakamura K, Katou H, Ishikawa S, Ito M, Honma T, Miyazaki N, Takehisa K, Sano S, Matsumoto S, Suda A, Baba K, Kawasaki A, Yamaguchi N, Akahane I, Tomizawa M Arao T. Simultaneous decrease of arsenic and cadmium in rice *Oryza sativa L* plants cultivated under submerged field conditions by the application of iron. Soil Sci Plant Nutri. 2016;62:340–8.
32. Seyfferth AL, Webb SM, Andrews JC, Fendorf S. Arsenic localization, speciation, and co-occurrence with iron on rice (*Oryza sativa* L.) roots having variable Fe coatings. Environ Sci Technol. 2010;44:8108–13.
33. Yamaguchi N, Ohkura T, Takahashi Y, Maejima Y, Arao T. Arsenic distribution and speciation near rice roots influenced by iron plaques and redox conditions of the soil matrix. Environ Sci Technol. 2014;48:1549–56.
34. Ma JF, Yamaji N, Mitani N, Xu XY, Su YH, McGrath SP, Zhao FJ. Transporters of arsenite in rice and their role in arsenic accumulation in rice grain. Proc Natl Acad Sci U S A. 2008;105:9931–5.
35. Seyfferth AL, Morris AH, Gill R, Kearns KA, Mann JN, Paukett M, Leskanic C. Soil incorporation of silica-rich rice husk decreases inorganic arsenic in rice grain. J Agric Food Chem. 2016;64:3760–6.
36. Suda A, Makino T. Effect of organic amendments on arsenic solubilization in soils during long-term flooded incubation. In J Environ Sci Technol. 2016;13:2375–82.
37. Ishikawa S, Ishimaru Y, Igura M, Kuramata M, Abe T, Senoura T, Hase Y, Arao T, Nishizawa NK, Nakanishi H. Ion-beam irradiation, gene identification, and marker-assisted breeding in the development of low-cadmium rice. Proc Natl Acad Sci U S A. 2012;109:19166–71.
38. Ishikawa S, Makino T, Ito M, Harada K, Nakada H, Nishida I, Nishimura M, Tokunaga T, Shirao K, Yoshizawa C, Matsuyama M, Abe T, Arao T. Low cadmium rice *Oryza sativa L* cultivar can simultaneously reduce arsenic and cadmium concentrations in rice grains. Soil Sci Plant Nutri. 2016;62:327–39.
39. Kuramata M, Abe T, Matsumoto S, Ishikawa S. Arsenic accumulation and speciation in Japanese paddy rice cultivars. Soil Sci Plant Nutri. 2011;57:248–58.

40. Song WY, Yamaki T, Yamaji N, Ko D, Jung KH, Fujii-Kashino M, An G, Martinoia E, Lee Y, Ma JF. A rice ABC transporter, OsABCC1, reduces arsenic accumulation in the grain. Proc Natl Acad Sci U S A. 2014;111:15699–704.
41. Chen Y, Moore KL, Miller AJ, McGrath SP, Ma JF, Zhao FJ. The role of nodes in arsenic storage and distribution in rice. J Exp Bot. 2015;66:3717–24.
42. Hayashi S, Kuramata M, Abe T, Takagi H, Ozawa K, Ishikawa S. Phytochelatin synthase OsPCS1 plays a crucial role in reducing arsenic levels in rice grains. Plant J. 2017;91(5):840–8. https://doi.org/10.1111/tpj.13612.
43. Duan GL, Kamiya T, Ishikawa S, Arao T, Fujiwara T. Expressing ScACR3 in rice enhanced arsenite efflux and reduced arsenic accumulation in rice grains. Plant Cell Physiol. 2012;53:154–63.
44. Clemens S, Ma JF. Toxic heavy metal and metalloid accumulation in crop plants and foods. Annu Rev Plant Biol. 2016;67:489–512.
45. Arao T, Kawasaki A, Baba K, Matsumoto S. Effects of arsenic compound amendment on arsenic speciation in rice grain. Environ Sci Technol. 2011;45:1291–7.
46. Li RY, Ago Y, Liu WJ, Mitani N, Feldmann J, McGrath SP, Ma JF, Zhao FJ. The rice aquaporin Lsi1 mediates uptake of methylated arsenic species. Plant Physiol. 2009;150:2071–80.
47. Carey AM, Scheckel KG, Lombi E, Newville M, Choi Y, Norton GJ, Charnock JM, Feldmann J, Price AH, Meharg AA. Grain unloading of arsenic species in rice. Plant Physiol. 2010;152:309–19.
48. Lomax C, Liu WJ, Wu L, Xue K, Xiong J, Zhou J, McGrath SP, Meharg AA, Miller AJ, Zhao FJ. Methylated arsenic species in plants originate from soil microorganisms. New Phytol. 2012;193:665–72.
49. Kuramata M, Sakakibara F, Kataoka R, Abe T, Asano M, Baba K, Takagi K Ishikawa S. Arsenic biotransformation by Streptomyces sp. isolated from rice rhizosphere. Environ Microbiol. 2015;17:1897–909.
50. Kuramata M, Abe T, Kawasaki A, Ebana K, Shibaya T, Yano M, Ishikawa S. Genetic diversity of arsenic accumulation in rice and QTL analysis of methylated arsenic in rice grains. Rice. 2013;6:3.

Chapter 11
Development of Low-Cadmium-Accumulating Rice

Satoru Ishikawa, Tadashi Abe, Masato Kuramata, and Shimpei Hayashi

Abstract Cadmium (Cd) has an important impact on agriculture, and the excessive consumption of Cd from contaminated food crops can lead to toxicity in humans. Rice (*Oryza sativa* L.) is the greatest source of dietary intake of Cd for populations that consume rice as a staple food. However, there is currently no practical technique designed to substantially reduce the Cd contamination of rice. Here, we report a new rice cultivar, Koshihikari Kan No. 1, which does not accumulate Cd in the grains. Koshihikari Kan No. 1 is a mutant produced from ion-beam irradiation. This cultivar has a single-nucleotide deletion in *OsNRAMP5*, which encodes a manganese transporter that incidentally transports Cd into rice; this deletion results in a decreased root Cd uptake. In Cd-contaminated paddy fields, Koshihikari Kan No. 1 showed nearly undetectable Cd concentrations in the grains and exhibited no economically adverse traits. A DNA marker, which detects the mutated region of *OsNRAMP5*, has been developed to facilitate marker-assisted breeding of cultivars with low-Cd traits. Our findings will help to greatly reduce Cd levels in paddy rice.

Keywords Cadmium · Ion beam · Marker-assisted breeding · OsNRAMP5 transporter · Rice mutant

S. Ishikawa (✉) · T. Abe · M. Kuramata
Institute for Agro Environmental Sciences, National Agriculture and Food Research Organization (NARO), Tsukuba, Ibaraki, Japan
e-mail: isatoru@affrc.go.jp

S. Hayashi
Institute for Agro-Environmental Sciences, National Agriculture and Food Research Organization (NARO), Tsukuba, Ibaraki, Japan

Institute of Agrobiological Sciences, NARO, Tsukuba, Ibaraki, Japan

S. Himeno, K. Aoshima (eds.), *Cadmium Toxicity*, Current Topics in Environmental Health and Preventive Medicine,
https://doi.org/10.1007/978-981-13-3630-0_11

11.1 Introduction

Cadmium (Cd) is a heavy metal element that is present in the soil naturally and is also provided by anthropogenic sources, including atmospheric deposition, fertilizers, manures, and other agricultural materials [1]. Cd accumulates in the human body for many years, leading to a potential risk of chronic toxicity with excessive intake of food containing Cd. In Japan, *itai-itai* disease (renal osteomalacia), which is characterized by complaints of spinal and leg bone pain, was recognized as a type of chronic toxicity induced by excess Cd contamination in drinking water and crops, including staple rice [2]. Since then, the contamination of rice has been monitored to prevent it from being distributed to consumers in Japan, in accordance with the Food Sanitation Act established in 1969. The Joint FAO/WHO Expert Committee on Food Additives established a provisional tolerable monthly Cd intake of 25 μg kg^{-1} body weight to reduce the risk of Cd poisoning [3], and the Codex Alimentarius Commission of the FAO/WHO determined maximum Cd levels in food crops [4]. The international criterion for rice Cd has been determined to be 0.4 mg kg^{-1} of polished rice. Rice is a staple food for nearly half of the world's population, and global production and consumption of rice increased by approximately 300% from 1960 to 2011 [5]. The demand for rice continues to grow, so it is necessary to produce low-Cd rice to reduce the potential risk that Cd poses to human health.

Keeping anaerobic flooding conditions is widely practiced in paddy fields that have the potential to produce rice contaminated with Cd in Japan. Flooding paddy fields shift soil redox potentials (Eh) toward a reduced state, where Cd ions precipitate with sulfide ions and form less-soluble cadmium sulfide (CdS), resulting in a decreased Cd uptake by rice [6]. However, the effects are unstable depending on soil properties, weather conditions, reservation of irrigation water, etc. Although soil dressing (i.e., replacement of topsoil) is proven as the most effective method to reduce Cd concentration in rice grains [6], this method is costly and has an environmental impact, because unpolluted soils are primarily obtained by cutting into mountains.

Selection and breeding of low-Cd-accumulating cultivars are the most cost-effective and environmentally friendly methods for reducing the contamination risk of Cd in crops [7]. There are substantial genotypic differences in Cd accumulation in rice [8, 9]. Generally, Cd concentrations are higher in *indica*-type cultivars than in *japonica*-type ones. The genetic loci that determine genotypic differences in rice Cd accumulation have been shown by the quantitative trait locus analysis for several mapping populations [10, 11]. In addition, the physiological and molecular processes of Cd transport in rice have been increasingly understood by successively identifying genes involved in root Cd uptake [12–16], root vacuole sequestration of Cd [17, 18], root xylem loading of Cd [19], and phloem transport of Cd in the node [20, 21]. However, since rice germplasm showing consistently lower grain Cd concentrations than Japanese elite cultivars has not been found, a breeding to confer low-Cd property into Japanese cultivars was less attainable. Mutations, induced by chemicals, such as ethyl methanesulfonate, or ionizing radiation, such as X-ray and gamma-ray, are a powerful way to explore novel mutants with favorable traits in agriculture. Among mutagens, energetic heavy-ion beams have been used to generate desirable mutants in higher plants; they can induce mutations with high frequency at a relative low dose at which virtually all plants survive, and they induce a broad spectrum of phenotypes without affecting other plant characteristics [22, 23]. In addition, mutants produced by ion-beam radiation are not transgenic plants and are therefore more likely to be accepted by consumers, and they are a practical choice for agriculture.

Using ion-beam mutagenesis, we constructed the rice mutant library and found three mutants that accumulated very low Cd concentration in grains. These mutants had different mutations of the *OsNRAMP5* gene, which encodes a manganese (Mn) transporter in rice [12]. On the basis of the differences in agronomic traits between a parent *Japonica* rice cultivar Koshihikari and mutants, we selected one mutant with the most resemblance to the Koshihikari cultivar. The mutant was recently accepted as a new cultivar (Koshihikari Kan No. 1, herein Kan1) by the Ministry of Agriculture, Forestry and Fisheries of Japan. In this chapter, we will provide the detailed properties of Kan1.

11.2 Mutation and Screening for Low-Cd Rice Mutants

The scheme to isolate the low-Cd rice mutants is illustrated in Fig. 11.1a. Husked seeds of Koshihikari (herein KSH) were irradiated with 320 MeV carbon ions ($^{12}C^{6+}$) from an azimuthally varying field cyclotron (Japan Atomic Energy Agency, Takasaki, Gunma, Japan) at a dose of 40 Gy. KSH is the most popular rice cultivar grown in Japan because of its good taste and eating quality, and it is grown in nearly 40% of the rice fields in Japan [24]. Approximately 4000 M_1 seeds were grown and self-pollinated, and the resultant seeds (M_2) were bulked. To isolate candidate mutants with low-grain-Cd concentrations, approximately 3000 M_2 seeds were transplanted into small plastic pots filled with Cd-contaminated paddy soil. The soil Cd concentration was 1.8 mg Cd kg^{-1} when determined by 0.1 M hydrochloric acid (HCl) extraction (the method stipulated by the *Agricultural Land-Soil Pollution Prevention Law* in Japan). KSH plants were also grown in Cd-contaminated paddy soil. All M_2 and KSH plants were submerged until the booting stage, and then water was withheld to increase the bioavailable Cd concentration in the soil and to enhance subsequent Cd uptake. After drainage, water was supplied from the top when required, and all plants were grown under aerobic soil conditions until grain maturity. The grains were harvested from all plants and analyzed to determine their Cd concentration by inductively coupled plasma mass spectrometry (ICP-MS).

Three low-Cd mutants (named temporarily as *lcd-kmt1*, *lcd-kmt2*, and *lcd-kmt3*) were isolated in the first screening. The grain Cd concentration in KSH averaged 1.73 mg kg^{-1}, whereas these mutants showed values <0.05 mg kg^{-1} (Fig. 11.1b). To characterize Cd uptake of these mutants at the vegetative stage, the seedlings were

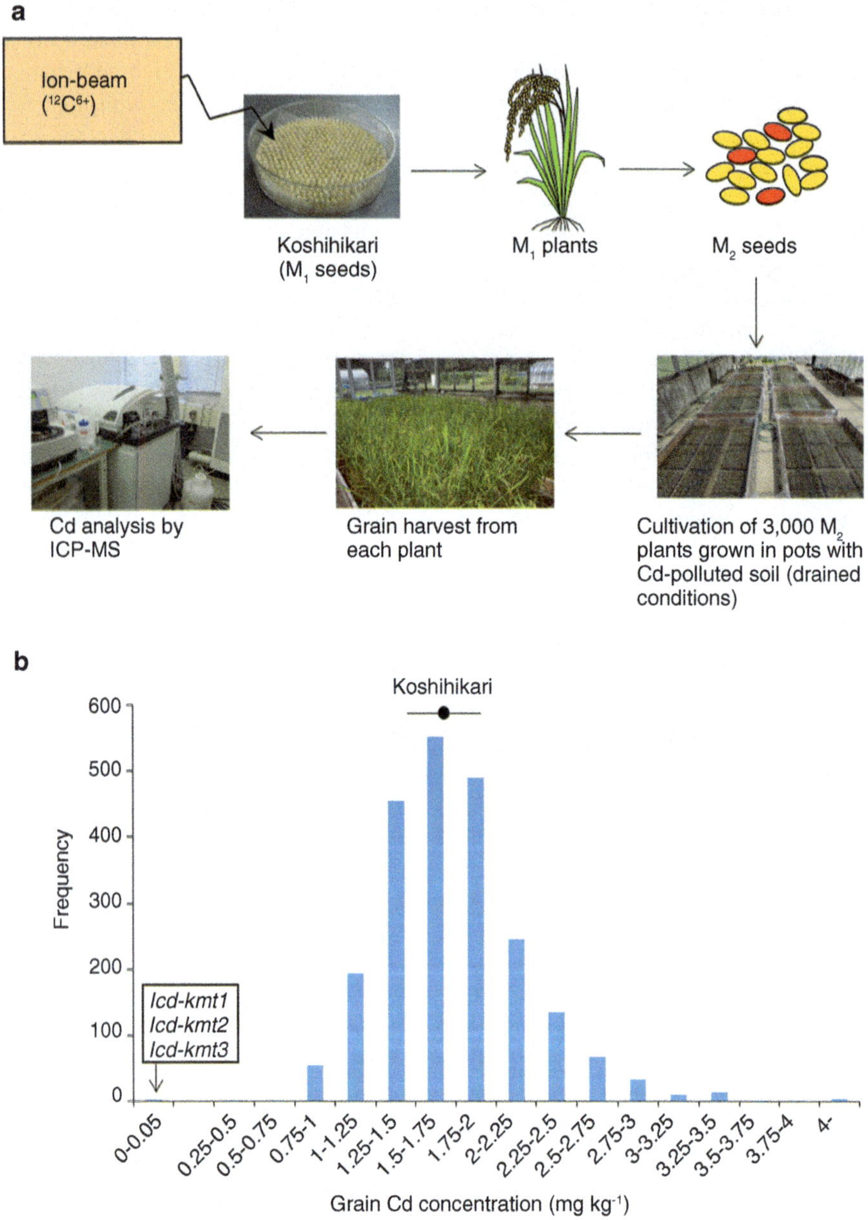

Fig. 11.1 (**a**) A schematic method to isolate low-Cd rice mutants after ion-beam irradiation. (**b**) Frequency distribution of grain Cd concentration in rice mutants grown in pots with Cd-polluted soil. The circle and range bar represent the mean and standard deviation (SD) of grain Cd concentration in Koshihikari. Three mutants, *lcd-kmt1*, *lcd-kmt2*, and *lcd-kmt3*, were isolated by means of (**a**)

exposed to Cd in hydroponics, and the Cd concentrations in roots and shoots were compared between the mutants and KSH. The results suggested that all mutants are characterized by a decreased root Cd uptake.

11.3 Field Performance

The mutant line *lcd-kmt2* was selected for registration of rice variety and was named Koshihikari Kan No. 1 (Kan1). To evaluate whether the low-Cd trait of Kan1 is exerted for field trails, the plants of KSH and Kan1 were cultivated in three Cd-polluted paddy fields in different regions of Japan. The soil Cd concentrations were 1.35 mg kg^{-1} (Field A), 1.21 mg kg^{-1} (Field B), and 0.35 mg kg^{-1} (Field C), and all soils were categorized as Gleysols. After 1 month of planting, the fields were managed by means of drainage and then irrigated intermittently until grain maturity. The grain and straw Cd concentrations in KSH and Kan1 were analyzed by ICP-MS, and the results were shown in Fig. 11.2a, b. The Cd concentrations of KSH grains

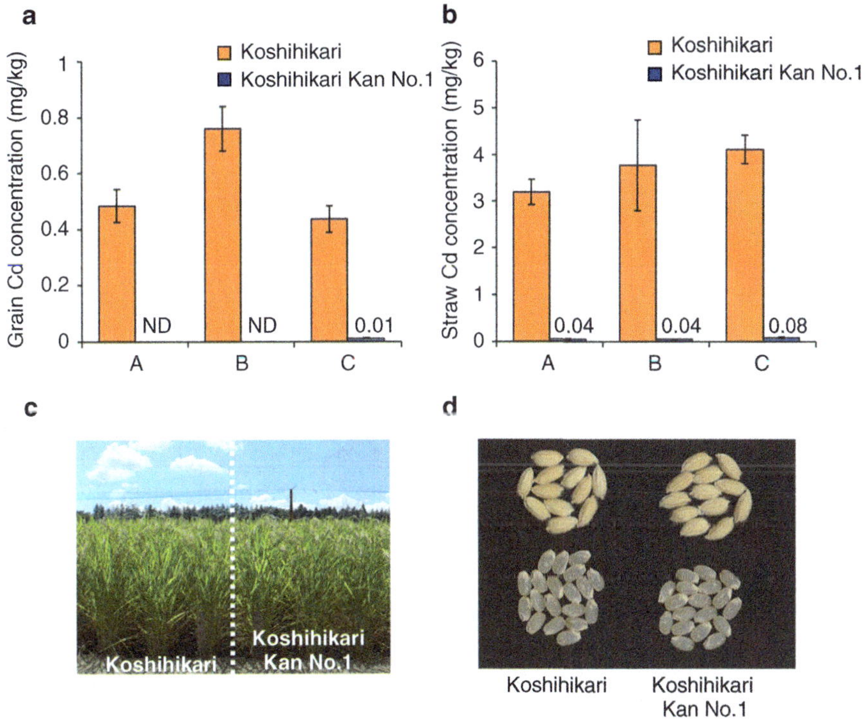

Fig. 11.2 (**a** and **b**) Grain and straw Cd concentrations in Koshihikari and Koshihikari Kan No. 1 grown in three Cd-polluted paddy fields in 2012. (**c**) Plant morphologies of Koshihikari and Koshihikari Kan No. 1 grown in the paddy field. (**d**) Morphologies of rice seeds and unpolished rice grains

Table 11.1 Comparisons of agronomic traits between Koshihikari and Koshihikari Kan No. 1

Cultivar	Days to heading	Culm length (cm)	Panicle length (cm)	Panicle number (m^2)	Grain yield (kg/a)	1000-grain weight (g)	Eating quality
Koshihikari	85	90	17.5	336	54.6	21.0	0.0
Koshihikari Kan No. 1	87	88	17.2	330	54.4	21.0	0.2

All data represent the average value of each cultivar grown in an uncontaminated field in 2011 and 2012

exceeded the maximum limit set by the Codex (0.4 mg kg^{-1}), whereas those of Kan1 grains were below the limit of quantification (<0.01 mg kg^{-1}) or 0.01 mg kg^{-1}. The straw Cd concentrations were also much lower in Kan1 than in the KSH. Thus, we confirmed the low-Cd property of Kan1 through the field tests.

The Kan1 was evaluated whether agronomic traits such as plant morphology and grain yield differed from the parent KSH. We did not find apparent difference in plant morphology between the Kan1 and KSH grown in the paddy field (Fig. 11.2c). The grain shape was similar between the two cultivars (Fig. 11.2d). The days to heading of Kan1 were delayed 2 days from that of KSH, although it was not a significant difference (Table 11.1). In addition, there were no significant differences in other agronomic traits such as culm length, panicle length, panicle number, grain yield, and 1000-grain weight. To evaluate eating quality of Kan1, we conducted the sensory test by a panel of 20 judges, who had been trained in the scoring of each component of eating quality. The judges evaluated five components, consisting of glossiness, smell, taste, stickiness, and hardness for cooked rice, with scores from −3.0 (extremely poor) to +3.0 (excellent), compared with the reference cultivar KSH (score 0.0). The average score from the 20 judges showed that there was not significant difference in eating quality between two cultivars, although the score of Kan1 was 0.2 points higher than that of KSH (Table 11.1). Thus, Kan1 exhibited no agriculturally or economically adverse traits.

Kan1 can contribute to a simultaneous decrease of Cd and arsenic (As) in rice grains. As in rice should be noted because it is a major source of inorganic As exposure in humans. In general, flooding increases the grain As concentrations, whereas drainage increases the Cd concentrations in rice. Thus, there is a trade-off between As and Cd uptake by rice depending on the type of water management. Two cultivars, KSH and Kan1, were cultivated in nine paddy fields with different soil properties under three water regimes: flooding conditions (FLD), alternate wetting and drying conditions (AWD), and water-saving conditions (WAS). The AWD and WAS regimes greatly reduced grain As concentrations relative to the FLD regime: the mean grain As concentrations of KSH and Kan1 were 27.3 and 27.4% less in the AWD and 48.2 and 43.1% less in the WAS, respectively, than in the FLD. Although AWD and WAS regimes remarkably increased grain Cd concentrations in KSH, Kan1 had nearly undetectable levels of grain Cd in all treatments. Therefore, growing the Kan1 cultivar aerobically is the most practical way to simultaneously reduce Cd and As contents in the rice grains [25].

11.4 Genetic Mechanisms

Genomic analysis based on map-based cloning revealed that Kan1 had a single-nucleotide deletion in exon IX of *OsNRAMP5*. The two other mutants, *lcd-kmt1* and *lcd-kmt3*, had different mutations of the same *OsNRAMP5*gene: a transposon insertion in the terminal portion of exon X for *lcd-kmt1* and large deletion including *OsNRAMP5* for *lcd-kmt3*. We registered two mutant genes as *osnramp5-1* for *lcd-kmt1* (accession no. AB690552) and *osnramp5-2* for Kan1 (*lcd-kmt2*) (AB690553) into the DNA Data Bank of Japan. *OsNRAMP5* (AB690551) of the KSH encoded a 538 amino acid (AA). The *osnramp5-1* open reading frame was completely translated, but an 11 AA of the KSH was replaced with 17 AA (an increase of 6 AA) at the terminal position of exon X, resulting in a 544 AA. The single deletion in *osnramp5-2* resulted in an aberrant translation of 53 AA before a new stop codon at AA 358.

Histochemical staining of β-glucuronidase (GUS) activity in transgenic rice showed that the *OsNRAMP5* promoter is expressed in all cell layers, with strong staining of the epidermis and stele in the elongating and mature zones of the roots [13]. To investigate subcellular localization of OsNRAMP5 and osnramp5-2, these proteins were C-terminally tagged with the green fluorescent protein and expressed in rice protoplasts. Green fluorescence of OsNRAMP5 was detected in covering the whole plasma membrane, whereas that of osnramp5-2 was scattered on the plasma membrane (Fig. 11.3a). We tested the influence of *OsNRAMP5* and *osnramp5-2* expression on Cd toxicity in the Cd-sensitive yeast mutant *Δycf1*. The growth of transformed yeast cells expressing *OsNRAMP5* was strongly impaired compared with yeast transformed with the empty vector control, whereas the growth of yeast cells expressing *osnramp5-2* was not influenced by the Cd treatment (Fig. 11.3b). Since OsNRAMP5 has been reported to be a Mn transporter in rice [15], we tested whether *OsNRAMP5* in KSH and *osnramp5-2* in Kan1 could complement yeast strains *Δsmf1*, which are defective in Mn^{2+} uptake. *OsNRAMP5* restored the growth of *Δsmf1*, whereas the growth of yeast expressing *osnramp5-2* was identical to that of the vector control (Fig. 11.3b). These data demonstrated that *OsNRAMP5* encodes a functional pro tein that mediates Cd and Mn, whereas the transporter encoded by *osnramp5-2* was defective in transport of these metals.

According to these results, we drew a schematic diagram of the genetic and molecular mechanism for the decreased Cd uptake in Kan1 (Fig. 11.3c). In KSH, the roots can absorb Cd through the functional OsNRAMP5 protein. On the other hand, Kan1 is not able to absorb Cd, probably because of the alteration in protein structure of OsNRAMP5. If, for Kan1, a small amount of Cd enters the root cells via other metal transporters such as OsIRT1 [14] and OsNRAMP1 [16], a kind of "firewall" system might sequester Cd in the root vacuoles via a functional OsHMA3 transporter [18]. Therefore, a defective gene (*osnramp5-2*) working together with a functional gene (*OsHMA3*) may be responsible for the drastic decrease in grain and straw Cd concentration in the Kan1 cultivar.

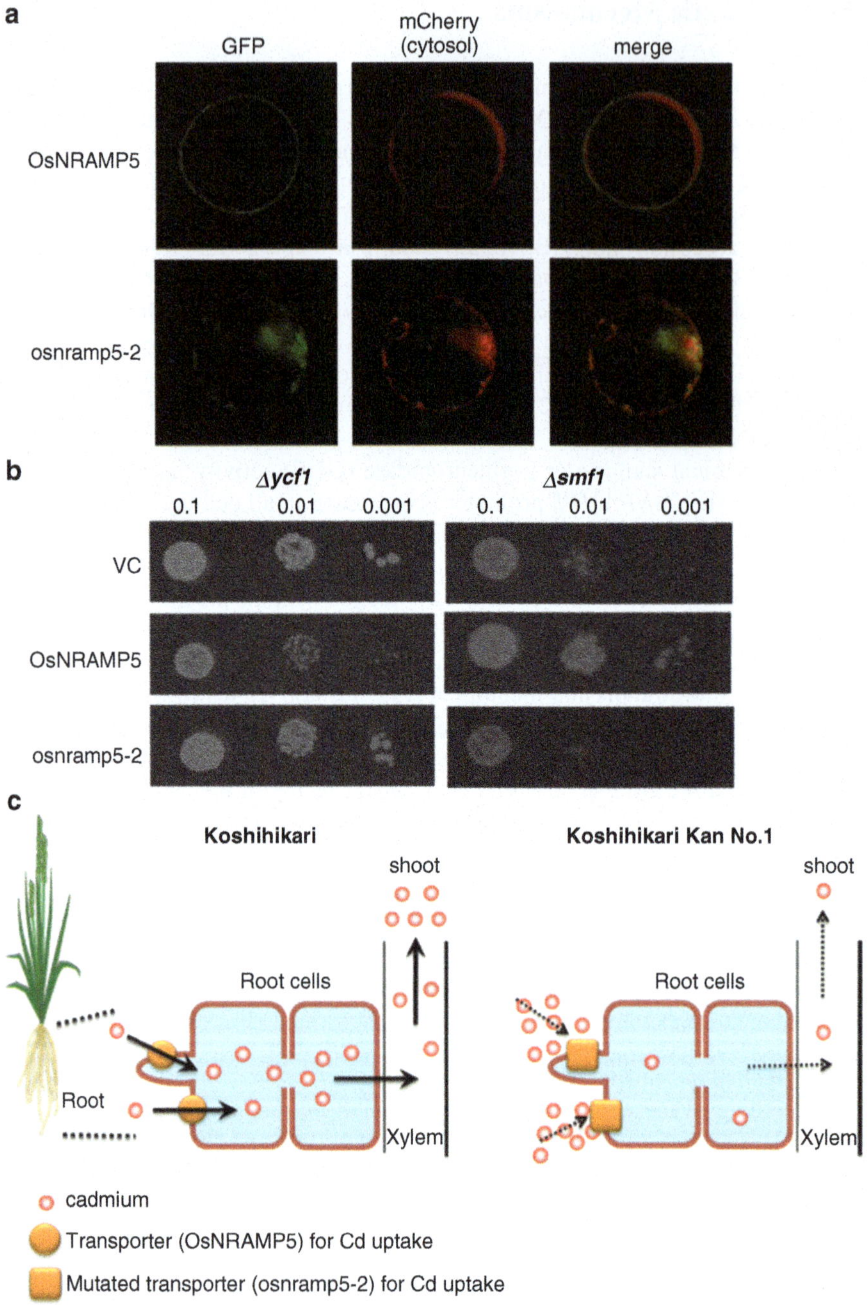

Fig. 11.3 (**a**) Subcellular localization of OsNRAMP5 and osnramp5-2 tagged with the green fluorescent protein in rice protoplasts. (**b**) Growth of *Δycf1* (Cd-sensitive mutant) and *Δsmf1* (Mn-uptake-defective mutant) yeast cells expressing *OsNRAMP5* or *osnramp5-2*. (**c**) Schematic diagram of the genetic and molecular mechanism for decreased Cd uptake in Koshihikari Kan No. 1

11.5 DNA Marker-Assisted Breeding for New Low-Cd Cultivars

DNA markers that detect polymorphism in the region of *osnramp5-2* can be used to develop new cultivars with the low-Cd trait. Because genomic sequences between *osnramp5-2* and *OsNRAMP5* differ in length by only 1 bp, we designed a specific primer set to amplify the mutated region, followed by digestion of the resulting PCR products with a restriction enzyme. The *Fsp*I enzyme cut the PCR product of Kan1 into two fragments of equal size, whereas the PCR product of KSH was not cut by this enzyme, resulting in different sizes of PCR products between KSH and Kan1, which can be observed on agarose gel. We illustrated a scheme of marker-assisted breeding to develop new cultivars with low-Cd trait (Fig. 11.4). The procedure is as follows: Kan1 is crossed with "cultivar A," a recurrent parent, and the resultant F_1 plant is backcrossed to "cultivar A," checking for the *osnramp5-2* allele using the DNA marker. After several backcrosses, the resultant BC_nF_1 is self-pollinated, and the BC_nF_2 with the homozygous allele for *osnramp5-2*, in the genetic background of "cultivar A," is selected. Since 2012, breeding programs have been implemented to transfer the *osnramp5-2* allele into more than 100 rice cultivars, including promising lines in Japan.

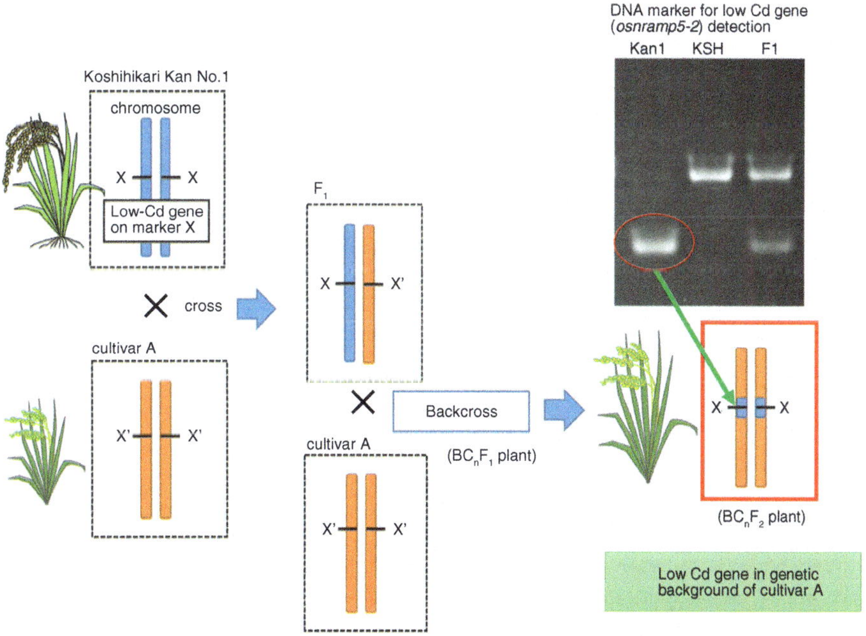

Fig. 11.4 Flow chart of marker-assisted breeding to develop new cultivars with low-Cd traits

11.6 Important Notes for Kan1 Cultivation

Although no significant difference in plant growth between Kan1 and KSH grown in paddy fields was observed as presented above (Fig. 11.2c), we should pay attention to Mn concentrations in Kan1 because Kan1 is defective in root Mn uptake through the OsNRAMP5 protein. In previous studies, the Mn concentrations in grains and straws of Kan1 showed an order of magnitude lower than those of KSH, depending on exchangeable Mn concentrations in soils and water regimes [12, 25]. Paddy rice has a good system to adapt for wide changes in Mn levels in the environment [26]. Under high-Mn conditions, the OsNRAMP3 protein, which acts as a Mn transporter at nodes, is rapidly degraded, and a high amount of Mn stayed in old tissues. Under low-Mn condition, Mn is preferentially delivered to young leaves and panicles via OsNRAMP3. Thus, we speculate that Kan1 might use an efficiently small amount of Mn for normal growth.

It has been reported that Kan1 was more susceptible to brown spot disease, induced by a filamentous fungus (*Cochliobolus miyabeanus*), than KSH when they were cultivated in paddy soils with low concentrations of K, Ca, Mg, Si, Fe, and Mn [27]. The brown spot disease negatively affects yield and grain quality. The pathogenesis rate can be governed by the physiological condition of the rice plant, and deficiencies of Si and Mn increase rice susceptibility to brown spot [28]. By applying Mn-bearing material to the paddy soils, the pathogenesis rate of brown spots in Kan1 was significantly decreased with the increased grain yield. Therefore, we recommend an Mn application for Kan1 grown in low-Mn conditions.

References

1. Alloway BJ, Steinnes E. Anthropogenic additions of cadmium to soils. In: McLaughlin MJ, Singh BR, editors. Cadmium in soils and plants. The Netherlands: Kluwer Academic Publishers; 1999. p. 97–123.
2. Tsuchiya K. Epidemiological-studies on cadmium in environment in Japan - etiology of itai-itai disease. Fed Proc. 1976;35:2412–8.
3. FAO/WHO. Joint FAO/WHO Expert Committee on Food Additives, Seventy-third Meeting, Geneva, 8–17 June 2010. Summary and Conclusions; 2010. http://www.who.int/foodsafety/publications/chem/summary73.pdf.
4. CODEX STAN 193-1995. Codex general standard for contaminants and toxins in foods and feed; 2008. www.fao.org/fileadmin/user_upload/agns/pdf/CXS_193e.pdf.
5. IRRI. World Production and Consumption of Domestic Milled Rice; 2011. http://ricestat.irri.org/vis/wrs_quickCharts.php
6. Arao T, Ishikawa S, Murakami M, Abe K, Maejima Y, Makino T. Heavy metal contamination of agricultural soil and countermeasures in Japan. Paddy Water Environ. 2010;8:247–57.
7. Grant CA, Clarke JM, Duguid S, Chaney RL. Selection and breeding of plant cultivars to minimize cadmium accumulation. Sci Total Environ. 2008;390:301–10.
8. Arao T, Ae N. Genotypic variations in cadmium levels of rice grain. Soil Sci Plant Nutr. 2003;49:473–9.

9. Uraguchi S, Mori S, Kuramata M, Kawasaki A, Arao T, Ishikawa S. Root-to-shoot Cd translocation via the xylem is the major process determining shoot and grain cadmium accumulation in rice. J Exp Bot. 2009;60:2677–88.
10. Ishikawa S, Abe T, Kuramata M, Yamaguchi M, Ando T, Yamamoto T, Yano M. A major quantitative trait locus for increasing cadmium-specific concentration in rice grain is located on the short arm of chromosome 7. J Exp Bot. 2010;61:923–34.
11. Ishikawa S, Ae N, Yano M. Chromosomal regions with quantitative trait loci controlling cadmium concentration in brown rice (*Oryza sativa*). New Phytol. 2005;168:345–50.
12. Ishikawa S, Ishimaru Y, Igura M, Kuramata M, Abe T, Senoura T, Hase Y, Arao T, Nishizawa NK, Nakanishi H. Ion-beam irradiation, gene identification, and marker-assisted breeding in the development of low-cadmium rice. Proc Natl Acad Sci U S A. 2012;109:19166–71.
13. Ishimaru Y, Takahashi R, Bashir K, Shimo H, Senoura T, Sugimoto K, Ono K, Yano M, Ishikawa S, Arao T, Nakanishi H, Nishizawa NK. Characterizing the role of rice NRAMP5 in manganese, iron and cadmium transport. Sci Rep. 2012;2:286.
14. Nakanishi H, Ogawa I, Ishimaru Y, Mori S, Nishizawa NK. Iron deficiency enhances cadmium uptake and translocation mediated by the Fe^{2+} transporters OsIRT1 and OsIRT2 in rice. Soil Sci Plant Nutr. 2006;52:464–9.
15. Sasaki A, Yamaji N, Yokosho K, Ma JF. Nramp5 is a major transporter responsible for manganese and cadmium uptake in rice. Plant Cell. 2012;24:2155–67.
16. Takahashi R, Ishimaru Y, Senoura T, Shimo H, Ishikawa S, Arao T, Nakanishi H, Nishizawa NK. The OsNRAMP1 iron transporter is involved in Cd accumulation in rice. J Exp Bot. 2011;62:4843–50.
17. Miyadate H, Adachi S, Hiraizumi A, Tezuka K, Nakazawa N, Kawamoto T, Katou K, Kodama I, Sakurai K, Takahashi H, Satoh-Nagasawa N, Watanabe A, Fujimura T, Akagi H. OsHMA3, a P-1B-type of ATPase affects root-to-shoot cadmium translocation in rice by mediating efflux into vacuoles. New Phytol. 2011;189:190–9.
18. Ueno D, Yamaji N, Kono I, Huang CF, Ando T, Yano M, Ma JF. Gene limiting cadmium accumulation in rice. Proc Natl Acad Sci U S A. 2010;107:16500–5.
19. Satoh-Nagasawa N, Mori M, Nakazawa N, Kawamoto T, Nagato Y, Sakurai K, Takahashi H, Watanabe A, Akagi H. Mutations in Rice (*Oryza sativa*) heavy metal ATPase 2 (OsHMA2) restrict the translocation of zinc and cadmium. Plant Cell Physiol. 2012;53:213–24.
20. Uraguchi S, Kamiya T, Sakamoto T, Kasai K, Sato Y, Nagamura Y, Yoshida A, Kyozuka J, Ishikawa S, Fujiwara T. Low-affinity cation transporter (OsLCT1) regulates cadmium transport into rice grains. Proc Natl Acad Sci U S A. 2011;108:20959–64.
21. Yamaji N, Xia JX, Mitani-Ueno N, Yokosho K, Ma JF. Preferential delivery of zinc to developing tissues in rice is mediated by P-type heavy metal ATPase OsHMA2. Plant Physiol. 2013b;162:927–39.
22. Kazama Y, Hirano T, Saito H, Liu Y, Ohbu S, Hayashi Y, Abe T. Characterization of highly efficient heavy-ion mutagenesis in Arabidopsis thaliana. BMC Plant Biol. 2011;11:161.
23. Tanaka A, Shikazono N, Hase Y. Studies on biological effects of ion beams on lethality, molecular nature of mutation, mutation rate, and spectrum of mutation phenotype for mutation breeding in higher plants. J Radiat Res. 2010;51:223–33.
24. Takeuchi Y, Hori K, Suzuki K, Nonoue Y, Takemoto-Kuno Y, Maeda H, Sato H, Hirabayashi H, Ohta H, Ishii T, Kato H, Nemoto H, Imbe T, Ohtsubo K, Yano M, Ando I. Major QTLs for eating quality of an elite Japanese rice cultivar, Koshihikari, on the short arm of chromosome 3. Breed Sci. 2008;58:437–45.
25. Ishikawa S, Makino T, Ito M, Harada K, Nakada H, Nishida I, Nishimura M, Tokunaga T, Shirao K, Yoshizawa C, Matsuyama M, Abe T, Arao T. Low-cadmium rice (*Oryza sativa* L.) cultivar can simultaneously reduce arsenic and cadmium concentrations in rice grains. Soil Sci Plant Nutr. 2016;62:327–39.
26. Yamaji N, Sasaki A, Xia JX, Yokosho K, Ma JF. A node-based switch for preferential distribution of manganese in rice. Nat Commun. 2013a;4:2442.

27. Honma T, Shiratori Y, Ohba H, Tsuchida T, Makino T, Abe T, Ishikawa S. Concentrations of nutrient content in rice variety KoshihikariKan no.1 and risk estimation of incidence of brown spot disease in different paddy fields. Jpn J Soil Sci Plant Nutr. 2017;88:213–20. (Japanense with English summary)
28. Junior LAZ, Rodrigues FA, Fontes RL, Korndorfer GH, Neves JC. Rice resistance to Brown spot mediated by silicon and its interaction with manganese. J Phytopathol. 2009;157:73–8.

Part IV
New Aspects in the Mechanisms of Cadmium Toxicity

Chapter 12
Novel Mechanisms of Cadmium-Induced Toxicity in Renal Cells

Jin-Yong Lee, Maki Tokumoto, and Masahiko Satoh

Abstract Cadmium (Cd) is an important industrial agent but is also an environmental pollutant that causes kidney disease. Chronic exposure to Cd results in its accumulation in proximal tubular cells of the kidney. This causes a variety of toxic effects that result in renal cell death. Current evidence indicates early phases of Cd-induced renal injury; for example, before renal cell death, Cd induces disruption to the transcription of apoptotic-related genes, dysregulation of autophagy, disruption of cell junction protein complexes, and accumulation of polyubiquitinated proteins. In this review, we discuss novel factors involved in Cd renal toxicity. This review may provide new insights for further elucidation of the mechanisms underlying Cd-induced toxicity in renal cells.

Keywords Cadmium · Kidney injury · Apoptosis · Autophagy · Cell-cell junction Ubiquitination

12.1 Introduction

Cadmium (Cd) is recognized as a health hazard, both through occupational and general exposure [1, 2]. There is widespread, low-level Cd contamination in agricultural soil in broad regions of the world [1]. The toxic effects of Cd due to occupational exposure were initially considered to cause lung and kidney damage, while dietary exposure caused kidney damage, osteomalacia, and fractures [3]. Because Cd has a long biological half-life (10–30 years), chronic exposure causes excessive accumulation in the body, mainly in the kidney, and overaccumulation of Cd in the kidney causes renal dysfunction [2]. Various studies have demonstrated that Cd-induced renal toxicity arises through several mechanisms, including oxidative stress, DNA damage, and apoptosis [4–6]. In addition to these classical mechanisms

J.-Y. Lee · M. Tokumoto · M. Satoh (✉)
Laboratory of Pharmaceutical Health Sciences, School of Pharmacy, Aichi Gakuin University, Nagoya, Aichi, Japan
e-mail: masahiko@dpc.agu.ac.jp

S. Himeno, K. Aoshima (eds.), *Cadmium Toxicity*, Current Topics in Environmental Health and Preventive Medicine,
https://doi.org/10.1007/978-981-13-3630-0_12

of toxicity, newly described pathways, involving novel targets, such as BIRC3 (Baculoviral IAP repeat containing 3), UBB, and Notch, are emerging [7–9].

In this review, we describe novel mechanisms of Cd toxicity, which have been elucidated using cultured kidney-derived cells and the rodent kidney. Although several recent reviews illustrate numerous Cd toxicity pathways, we focus on targets recently identified to be involved in Cd-induced apoptosis, dysregulation of autophagy, cell-cell junction disruption, and accumulation of ubiquitinated proteins. Further, we present our own perspectives on Cd-induced apoptosis initiated by transcriptional dysregulation of genes involved in the ubiquitin-proteasome system (UPS).

12.2 Apoptosis-Related Cd Toxicity in Renal Cells

Cd-induced renal toxicity is associated with apoptosis in proximal tubular cells. Pathways such as the mitochondria- and caspase-mediated pathway, endoplasmic reticulum (ER) stress-mediated unfolded protein response (UPR) signals, and calcium (Ca) release from the ER *via* G protein-coupled receptor phospholipase C (GPCR-PLC) activation are involved in Cd-induced apoptosis [10]. The apoptosis-inducing protein, p53, is a regulation factor of Cd-induced apoptosis in several cell lines [11, 12]. Our recent studies have shown that overaccumulation of p53 is associated with Cd-induced apoptosis in proximal tubular cells [13, 14]. Cd triggers the accumulation of p53 by inhibiting its degradation [13, 14]. p53 is immediately degraded by the UPS in normal conditions; however, Cd suppresses expression of *UBE2D* (ubiquitin-conjugating enzyme E2D) family genes, which encode the ubiquitin-conjugating (E2) enzyme involved in the ubiquitination of p53 [13, 14]. Interestingly, Cd did not affect proteasome activity [14]. In human proximal tubular cells, Cd inhibits *UBE2D2* and *UBE2D4* gene expression through the suppression of Yin Yang 1 (YY1) and forkhead box F1 (FOXF1) transcription activity [13]. The expression of apoptosis-related genes, *BAX* and *PUMA*, which are downstream of p53, is upregulated by Cd treatment. Meanwhile, knockdown of p53 attenuated the expression of *BAX* and *PUMA* both with and without Cd [13]. Notably, the modification of p53 by Cd may vary depending on cell type, as seen using rat small intestine epithelial cells, human brain microvascular endothelial cells, and human brain astrocytes [15]. Furthermore, Cd induces apoptosis by decreasing the level of BIRC3 in human proximal tubular cells. BIRC3 inhibits apoptosis by reducing the activity of caspases [9]. Cd decreases the level of BIRC3 protein by inhibiting *BIRC3* transcription [9]. Moreover, Cd-induced suppression of the aryl hydrocarbon receptor nuclear translocator (ARNT) transcription factor is involved in the inhibition of *BIRC3* gene expression. Knockdown of *BIRC3* as well as Cd treatment increases the activated form of caspase3 [9]. From our recent studies, we propose a new Cd-induced apoptosis pathway initiated by the suppression of transcription factors, such as YY1, FOXF1, and ARNT (Fig. 12.1). As a result, expression of *UBE2D* family genes and *BIRC3* is decreased, leading to increased levels of p53 and p53 downstream factors and activated caspase3 (Fig. 12.1).

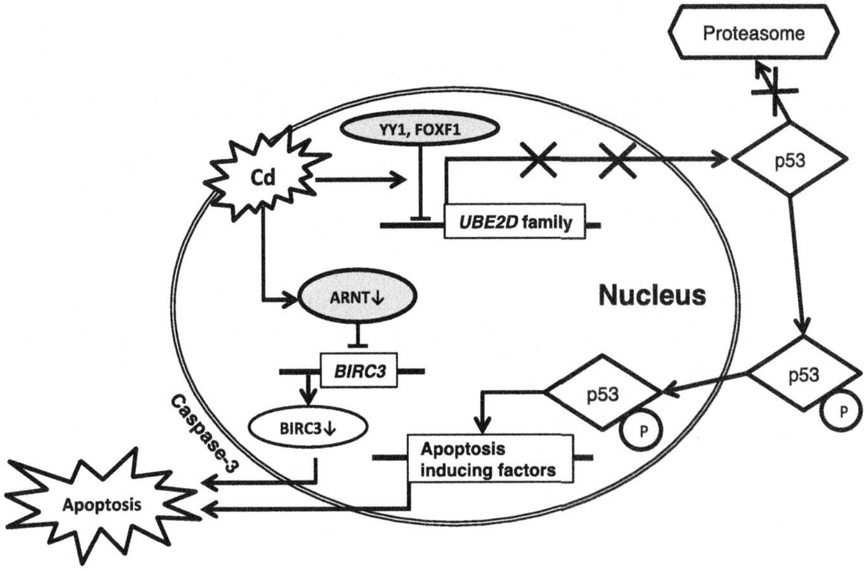

Fig. 12.1 Apoptosis-related Cd toxicity in renal cells. Cd suppresses the activity of YY1, FOXF1, and ARNT transcription factors. The suppressed transcription activity decreases the expression of *UBE2D* family genes and *BIRC3*. Downregulation of *UBE2D* family genes induces the accumulation of p53. Phosphorylated p53 in the nucleus promotes the expression of apoptosis-related genes. Moreover, downregulation of *BIRC3* expression by inhibition of ARNT activates caspase-3

12.3 Cd Toxicity in Renal Cells *via* Autophagy

Autophagy is an evolutionarily conserved "self-devouring" system for the destruction of cells [16–18]. Autophagy is a multistep process that is important to the degradation and recycling of intracellular components, damaged organelles, and misfolded proteins [16–18]. Autophagy acts as a cytoprotective mechanism in maintaining cellular homeostasis and adapting to environmental stress [16–18]; however, aberrant autophagy has been implicated in acute kidney injury and chronic kidney disease [19]. Recently, dysregulation of autophagy has been associated with Cd renal toxicity. The ER triggers autophagy to dispose of unfolded proteins in cells [20]. Intraperitoneal injection of 1 ppm or more of Cd for 1 week induces kidney cell autophagy in the mice [21]. Cd increases the level of cyclooxygenase2 (COX2) protein and expression of its encoding gene (*PTGS2*). COX2 inhibits a negative regulator of autophagy, mTOR (mammalian target of rapamycin) in the mouse kidney [21]. The Cd-increased COX2 level is mediated by the ER stress eukaryotic initiation factor 2α (eIF2α)-activating transcription factor 4 (ATF4) pathway in the mouse kidney and human embryonic kidney cells [21]. These results suggest that eIF2α-ATF4 pathway-mediated COX2 induction regulates Cd-induced autophagy. The treatment of mesangial cells (MES-13) with Cd induces autophagy and elevates cytosolic Ca^{2+} levels originating from the ER [22]. Autophagy is regulated by the activation of MAPKs (mitogen-activated protein kinases), such as ERK (extracellular signal-regulated kinase), JNK (Jun N-terminal kinase), and p38 [22–24]. Cd activates ERK, while inhibition of ERK suppresses Cd-induced autophagy [22]. In addition, a chelator of cytosolic Ca^{2+} reduced Cd-induced activation of ERK [22]. Therefore, Cd-induced autophagy is mediated by Ca^{2+}-ERK signaling. Autophagy includes an autophagosome formation process in which misfolded proteins or damaged organelles are first packaged by a double-membrane vesicle. Autophagosomes are then fused with lysosomes to form autolysosomes, inside which the packaged contents are degraded by lysosomal enzymes [25, 26]. Rab7, a member of the Rab family, is recruited to autophagosomes to promote lysosome fusion [27, 28]. Cd inhibits the recruitment of Rab7 to autophagosomes in rat proximal tubular (rPT) cells [25]. Cytosolic Ca^{2+} chelation not only suppresses autophagosome accumulation but also restores Cd-induced inhibition of Rab7 recruitment to autophagosomes in rPT cells [25]. These results suggest that Cd dysregulates autophagy by Ca^{2+} release-dependent alteration of Rab7 localization. Glycogen synthase kinase-3β (GSK-3β) is important for cell survival [29]. GSK-3β is activated by dephosphorylation at Ser9 [30]. Cd induces not only autophagy but also dephosphorylation of GSK-3β in MES-13 cells [31]. GSK-3β inhibition and genetic knockdown result in a reduction of Cd-induced autophagy [31]. GSK-3β is activated by reactive oxygen species (ROS) [32]. The ROS scavenger, *N*-acetylcysteine (NAC), reverses the dephosphorylation of GSK-3β induced by Cd. Moreover, NAC reduces autophagy induced by Cd [31] indicating that ROS-activated GSK-3β is crucial for Cd-induced autophagy. These studies identify GSK-3β, COX-2, ERK, and Rab7 as key factors regulating autophagy-dependent renal cell death (Fig. 12.2).

12.4 Damage to Renal Cell-Cell Adhesion by Cd

Recent studies on Cd toxicity in proximal tubules have revealed that Cd has a damaging effect on cell-cell junctions. A main element of this damage is disruption to cadherin-mediated adhesion [7, 33, 34]. Cadherins are cell adhesion molecules consisting of over 40 families, such as *E*-, *N*-, and *P*-cadherins [34, 35].Cadherins are transmembrane proteins localized at junctional complexes of epithelial cells. The extracellular domain of cadherins contains Ca^{2+}-binding sites. The intracellular domain is linked to β-catenin and α-catenin, which, in turn, links to the actin cytoskeleton [34, 36, 37]. *E*-cadherin is the predominant cadherin present in renal tubular epithelial cells. Recently, however, *N*-cadherin has also been detected in proximal tubular cells [34, 38]. In normal rat kidney, cadherins are localized in proximal tubular epithelial cells in a weak and diffuse pattern near the basolateral surface and in a more concentrated pattern over the apical surface. In kidney sections from rats subcutaneously injected with 0.6 ppm Cd (5 days per week for 6 months), cadherins were no longer localized along the apical or basolateral surface of proximal tubular

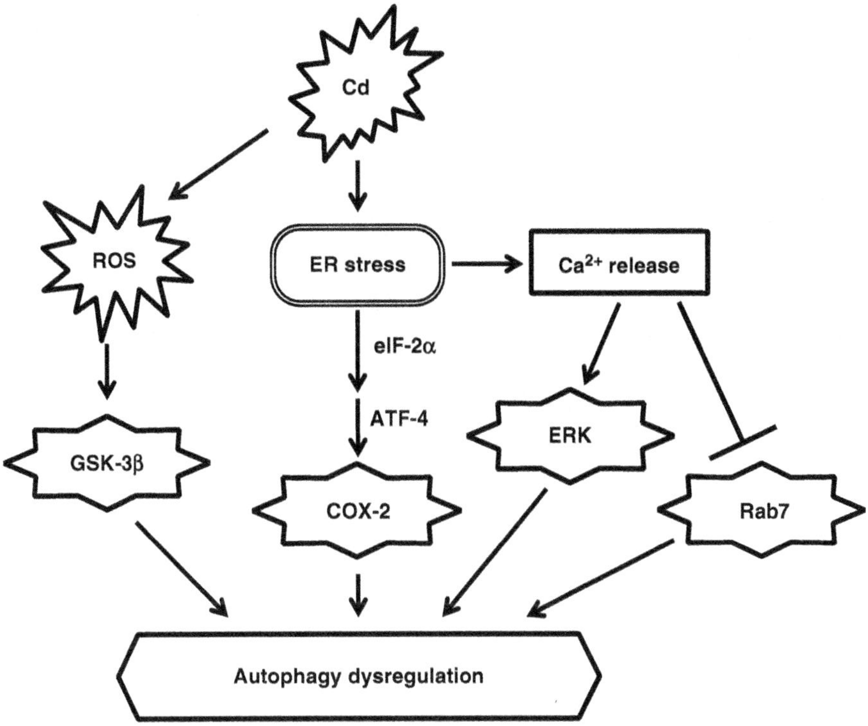

Fig. 12.2 Cd toxicity in renal cells *via* autophagy. Cd induces ER stress, which releases Ca^{2+} into the cytosol. The elevated Ca^{2+} level not only activates ERK but also changes Rab7 localization. ER stress induced by Cd increases COX2 levels *via* the eIF2α-ATF4 pathway. Cd increases ROS levels, which induces GSK-3β. Upregulated ERK, COX2, and GSK-3β and dysregulated Rab7 enhance autophagy dysregulation

epithelial cells but were diffusely present throughout the cytoplasm [34]. Cd-induced disruption of cadherin-catenin adhesion junctions was demonstrated in cultured rat kidney proximal tubule cells [33]. Cd decreased the protein level of *E*-cadherin; however, Cd increased the protein level of the zinc finger transcription factor, Snail, a repressor of *E*-cadherin expression in HK-2 cells [7]. Knockdown of Snail protected HK-2 cells from a Cd-induced reduction of *E*-cadherin expression and cellular damage [7]. Interestingly, Cd-induced Snail activation is associated with translocation of Notch intracellular domain (NICD) into the nucleus [7]. The Notch ligands, Jagged1 and Jagged2, are involved in Cd-induced release of NICD from the Notch complex [7]. The translocation of NICD into the nucleus is modulated by p53 activation; moreover, nuclear-translocated NICD activates PI3K/Akt signaling and results in Snail expression [7]. Therefore, Cd induced the disruption of cell-cell junctions through the translocation of cadherin and also by decreasing the expression of cadherin (Fig. 12.3).

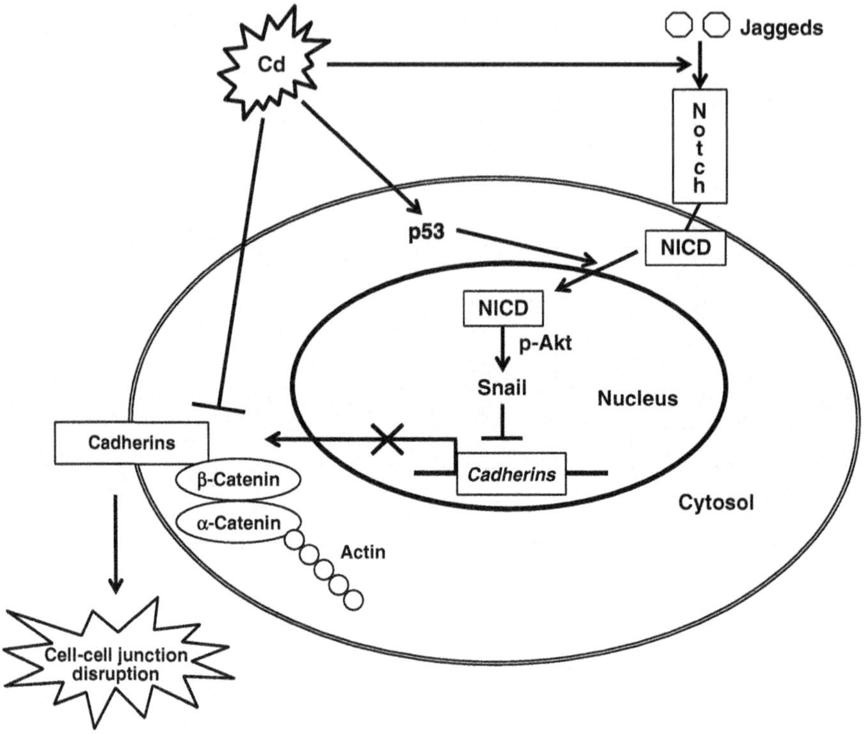

Fig. 12.3 Damage to renal cell-cell adhesion by Cd. Cd triggers binding of the ligands, Jagged1 and Jagged2, to Notch receptor. The Notch intracellular domain (NICD) localizes to the nucleus; Cd-activated p53 induces the localization of NICD to the nucleus. Nuclear NICD induces the expression of the transcription factor, Snail. Elevated Snail suppresses cadherin genes. In addition, Cd not only causes the mislocalization of cadherin but also inhibits the binding of β-catenin to cadherin. Aberrant cadherin function disrupts cell-cell adhesion junctions

12.5 The Ubiquitin-Proteasome System Related Pathway

Cd affects the UPS in several kidney cells through changes in UPS-related gene expression or accumulation of ubiquitinated proteins [8, 39–42]. Tokumoto et al. recently studied gene expression changes in normal rat proximal tubular epithelial cells (NRK-52E cells) exposed to Cd [41]. Among the genes whose expression was changed by Cd treatment, *Ube2d4*, which is involved in the UPS, was downregulated by Cd treatment [14, 41]. In the kidney of mice exposed to Cd for 12 months, several ubiquitination-related genes were downregulated [40]. The UPS is composed of ubiquitin-activating enzymes (E1), ubiquitin-conjugating enzymes (E2), ubiquitin ligases (E3), and ubiquitin-chain elongation factors (E4) [43]. A main event in the UPS is the conjugation of ubiquitin (Ub), a highly conserved 76 amino acid protein, to the substrate protein. Ub-coding genes are found

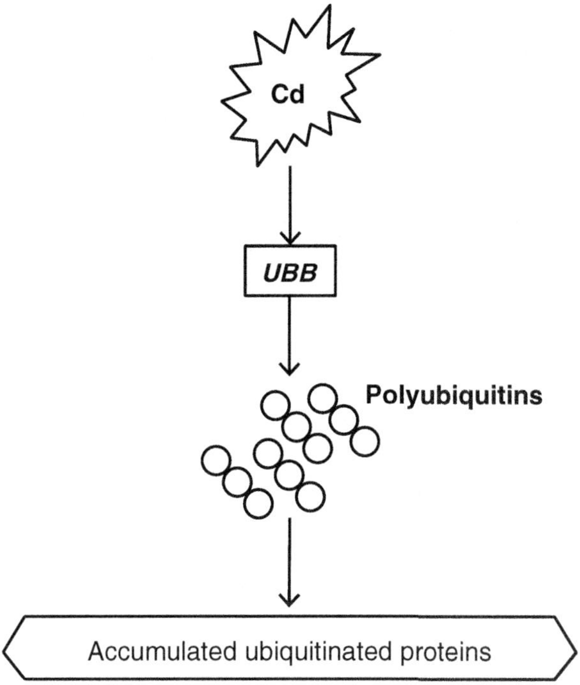

Fig. 12.4 Ubiquitin-proteasome system related pathway. Cd increases *UBB* gene expression prior to cytotoxicity. Increased *UBB* expression triggers polyubiquitination of proteins. Accumulated polyubiquitinated proteins are involved in Cd toxicity

in two forms. The first type, *UBB* and *UBC*, encodes Ub tandem repeats. The second type, *UBA52* and *UBA80*, encodes a monoUb fused to a tail protein [43, 44]. We demonstrated that prior to the appearance of Cd toxicity, *UBB*, *UBC*, and *UBA80* expression levels increase following Cd treatment in HK-2 cells [8]. Knockdown of *UBB* by siRNA significantly decreased Cd cytotoxicity [8]. Notably, Cd increased ubiquitinated protein levels in HK-2 cells, and knockdown of *UBB* blocked this process [8]. However, *UBA80* knockdown did not diminish Cd-induced protein ubiquitination [39]. Recent studies suggest that accumulation of ubiquitinated proteins is involved in Cd renal toxicity; moreover, polyubiquitin is preferably engaged in Cd-induced accumulation of ubiquitinated proteins in renal cells (Fig. 12.4).

12.6 Closing Remarks

In this review, we described several pathways through which Cd induces renal cell toxicity. We summarized four novel pathways: (1) a new apoptosis pathway initiated by the suppression of transcription factor activities, (2) novel factors mediating

Cd-induced autophagy dysregulation, (3) Cd disruption of cell-cell junctions through the alteration of cadherins, and (4) Cd-induced accumulation of ubiquitinated proteins mediated by increased polyubiquitin gene activity. Although other pathways have been reported using various cell lines, we focused on reports using kidney-derived cells and rodent kidney. Our previous reports have also suggested several new factors, such as *SLC39A1*, *GPRC5B*, and *HIS1H4C*, to be involved in Cd renal toxicity, although their precise mechanisms of action have not been elucidated [45–47]. These novel mechanisms may stimulate further studies into Cd renal toxicity.

Acknowledgments This work was supported by the Study of the Health Effects of Heavy Metals, organized by the Ministry of the Environment, Japan.

Conflict of Interest Statement The authors declare that they have no conflict of interest.

References

1. Akesson A, Barregard L, Bergdahl IA, Nordberg GF, Nordberg M, Skerfving S. Non-renal effects and the risk assessment of environmental cadmium exposure. Environ Health Perspect. 2014;122(5):431–8.
2. Jarup L, Akesson A. Current status of cadmium as an environmental health problem. Toxicol Appl Pharmacol. 2009;238(3):201–8.
3. Nordberg GF, Nogawa K, Nordberg M, Friberg LT. Cadmium. In: Nordberg GF, Fowler BA, Nordberg M, Friberg LT, editors. Handbook on the toxicology of metals. 3rd ed. Amsterdam and Boston: Academic Press/Elsevier; 2007. p. 445–86.
4. Bork U, Lee WK, Kuchler A, Dittmar T, Thevenod F. Cadmium-induced DNA damage triggers G(2)/M arrest via chk1/2 and cdc2 in p53-deficient kidney proximal tubule cells. Am J Physiol Renal Physiol. 2010;298(2):F255–65.
5. Gao D, Xu Z, Zhang X, Zhu C, Wang Y, Min W. Cadmium triggers kidney cell apoptosis of purse red common carp (Cyprinuscarpio) without caspase-8 activation. Dev Comp Immunol. 2013;41(4):728–37.
6. Nair AR, Lee WK, Smeets K, Swennen Q, Sanchez A, Thevenod F, Cuypers A. Glutathione and mitochondria determine acute defense responses and adaptive processes in cadmium-induced oxidative stress and toxicity of the kidney. Arch Toxicol. 2015;89(12):2273–89.
7. Fujiki K, Inamura H, Matsuoka M. Detrimental effects of Notch1 signaling activated by cadmium in renal proximal tubular epithelial cells. Cell Death Dis. 2014;5:e1378.
8. Lee JY, Tokumoto M, Fujiwara Y, Satoh M. Involvement of ubiquitin-coding genes in cadmium-induced protein ubiquitination in human proximal tubular cells. J Toxicol Sci. 2015;40(6):901–8.
9. Lee JY, Tokumoto M, Hwang GW, Lee MY, Satoh M. Identification of ARNT-regulated BIRC3 as the target factor in cadmium renal toxicity. Sci Rep. 2017;7:17287.
10. Fujiwara Y, Lee JY, Tokumoto M, Satoh M. Cadmium renal toxicity via apoptotic pathways. Biol Pharm Bull. 2012;35(11):1892–7.
11. Aimola P, Carmignani M, Volpe AR, Di Benedetto A, Claudio L, Waalkes MP, et al. Cadmium induces p53-dependent apoptosis in human prostate epithelial cells. PLoS One. 2012;7(3):e33647.
12. Son YO, Lee JC, Hitron JA, Pan J, Zhang Z, Shi X. Cadmium induces intracellular Ca^{2+}- and H_2O_2-dependent apoptosis through JNK- and p53-mediated pathways in skin epidermal cell line. Toxicol Sci. 2010;113(1):127–37.

13. Lee JY, Tokumoto M, Fujiwara Y, Hasegawa T, Seko Y, Shimada A, Satoh M. Accumulation of p53 via down-regulation of UBE2D family genes is a critical pathway for cadmium-induced renal toxicity. Sci Rep. 2016;6:21968.
14. Tokumoto M, Fujiwara Y, Shimada A, Hasegawa T, Seko Y, Nagase H, Satoh M. Cadmium toxicity is caused by accumulation of p53 through the down-regulation of Ube2d family genes in vitro and in vivo. J Toxicol Sci. 2011;36(2):191–200.
15. Lee JY, Tokumoto M, Hattori Y, Fujiwara Y, Shimada A, Satoh M. Different Regulation of p53 Expression by cadmium exposure in kidney, liver, intestine, vasculature, and brain astrocytes. Toxicol Res. 2016;32(1):73–80.
16. Mizushima N, Komatsu M. Autophagy: renovation of cells and tissues. Cell. 2011;147(4):728–41.
17. Mizushima N, Yoshimori T, Levine B. Methods in mammalian autophagy research. Cell. 2010;140(3):313–26.
18. Shintani T, Klionsky DJ. Autophagy in health and disease: a double-edged sword. Science. 2004;306:990–5.
19. Sureshbabu A, Ryter SW, Choi ME. Oxidative stress and autophagy: crucial modulators of kidney injury. Redox Biol. 2015;4:208–14.
20. Yorimitsu T, Nair U, Yang Z, Klionsky DJ. Endoplasmic reticulum stress triggers autophagy. J Biol Chem. 2006;281(40):30299–304.
21. Luo B, Lin Y, Jiang S, Huang L, Yao H, Zhuang Q, et al. Endoplasmic reticulum stress eIF2alpha-ATF4 pathway-mediated cyclooxygenase-2 induction regulates cadmium-induced autophagy in kidney. Cell Death Dis. 2016;7(6):e2251.
22. Wang SH, Shih YL, Ko WC, Wei YH, Shih CM. Cadmium-induced autophagy and apoptosis are mediated by a calcium signaling pathway. Cell Mol Life Sci. 2008;65(22):3640–52.
23. Ogier-Denis E, Pattingre S, El Benna J, Codogno P. Erk1/2-dependent phosphorylation of Galpha-interacting protein stimulates its GTPase accelerating activity and autophagy in human colon cancer cells. J Biol Chem. 2000;275(50):39090–5.
24. vom Dahl S, Dombrowski F, Schmitt M, Schliess F, Pfeifer U, Haussinger D. Cell hydration controls autophagosome formation in rat liver in a microtubule-dependent way downstream from p38MAPK activation. Biochem J. 2001;354(Pt 1):31–6.
25. Liu F, Wang XY, Zhou XP, Liu ZP, Song XB, Wang ZY, Wang L. Cadmium disrupts autophagic flux by inhibiting cytosolic Ca^{2+}-dependent autophagosome-lysosome fusion in primary rat proximal tubular cells. Toxicology. 2017;383:13–23.
26. Lu Y, Dong S, Hao B, Li C, Zhu K, Guo W, et al. Vacuolin-1 potently and reversibly inhibits autophagosome-lysosome fusion by activating RAB5A. Autophagy. 2014;10(11):1895–905.
27. Chua CE, Gan BQ, Tang BL. Involvement of members of the Rab family and related small GTPases in autophagosome formation and maturation. Cell Mol Life Sci. 2011;68(20):3349–58.
28. Gutierrez MG, Munafo DB, Beron W, Colombo MI. Rab7 is required for the normal progression of the autophagic pathway in mammalian cells. J Cell Sci. 2004;117(Pt 13):2687–97.
29. Hoeflich KP, Luo J, Rubie EA, Tsao MS, Jin O, Woodgett JR. Requirement for glycogen synthase kinase-3beta in cell survival and NF-kappaB activation. Nature. 2000;406(6791):86–90.
30. Forde JE, Dale TC. Glycogen synthase kinase 3: a key regulator of cellular fate. Cell Mol Life Sci. 2007;64(15):1930–44.
31. Wang SH, Shih YL, Kuo TC, Ko WC, Shih CM. Cadmium toxicity toward autophagy through ROS-activated GSK-3beta in mesangial cells. Toxicol Sci. 2009;108(1):124–31.
32. Pizarro JG, Yeste-Velasco M, Rimbau V, Casadesus G, Smith MA, Pallas M, et al. Neuroprotective effects of SB-415286 on hydrogen peroxide-induced cell death in B65 rat neuroblastoma cells and neurons. Int J Dev Neurosci. 2008;26(3–4):269–76.
33. Chakraborty PK, Lee WK, Molitor M, Wolff NA, Thevenod F. Cadmium induces Wnt signaling to upregulate proliferation and survival genes in sub-confluent kidney proximal tubule cells. Mol Cancer. 2010;9:102.
34. Prozialeck WC, Lamar PC, Lynch SM. Cadmium alters the localization of N-cadherin, E-cadherin, and beta-catenin in the proximal tubule epithelium. Toxicol Appl Pharmacol. 2003;189(3):180–95.

35. Nollet F, Kools P, van Roy F. Phylogenetic analysis of the cadherin superfamily allows identification of six major subfamilies besides several solitary members. J Mol Biol. 2000;299(3):551–72.
36. Adams CL, Nelson WJ. Cytomechanics of cadherin-mediated cell-cell adhesion. Curr Opin Cell Biol. 1998;10(5):572–7.
37. Lutz KL, Siahaan TJ. Molecular structure of the apical junction complex and its contribution to the paracellular barrier. J Pharm Sci. 1997;86(9):977–84.
38. Leussink BT, Litvinov SV, de Heer E, Slikkerveer A, van der Voet GB, Bruijn JA, de Wolff FA. Loss of homotypic epithelial cell adhesion by selective N-cadherin displacement in bismuth nephrotoxicity. Toxicol Appl Pharmacol. 2001;175(1):54–9.
39. Lee JY, Tokumoto M, Hwang GW, Satoh M. Cadmium-induced protein ubiquitination in UBA80 knockdown HK-2 cells. Fundam Toxicol Sci. 2016;3(6):281–4.
40. Tokumoto M, Lee JY, Fujiwara Y, Satoh M. DNA microarray expression analysis of mouse kidney following cadmium exposure for 12 months. J Toxicol Sci. 2013;38(5):799–802.
41. Tokumoto M, Ohtsu T, Honda A, Fujiwara Y, Nagase H, Satoh M. DNA microarray analysis of normal rat kidney epithelial cells treated with cadmium. J Toxicol Sci. 2011;36(1):127–9.
42. Uekusa H, Namimatsu M, Hiwatashi Y, Akimoto T, Nishida T, Takahashi S, Takahashi Y. Cadmium interferes with the degradation of ATF5 via a post-ubiquitination step of the proteasome degradation pathway. Biochem Biophys Res Commun. 2009;380(3):673–8.
43. Ravid T, Hochstrasser M. Diversity of degradation signals in the ubiquitin-proteasome system. Nat Rev Mol Cell Biol. 2008;9(9):679–90.
44. Baker RT, Board PG. The human ubiquitin-52 amino acid fusion protein gene shares several structural features with mammalian ribosomal protein genes. Nucleic Acids Res. 1991;19(5):1035–40.
45. Lee JY, Tokumoto M, Fujiwara Y, Lee MY, Satoh M. Effects of cadmium on the gene expression of SLC39A1 coding for ZIP1 protein. Fundam Toxicol Sci. 2014;1(4):131–3.
46. Lee JY, Tokumoto M, Fujiwara Y, Lee MY, Satoh M. The involvement of GPRC5B in cadmium toxicity in HK-2 cells. Fundam Toxicol Sci. 2014;1(4):165–7.
47. Lee JY, Tokumoto M, Satoh M. The enhancement effect of HIST1H4C knockdown on cadmium toxicity in human proximal tubular cells. Fundam Toxicol Sci. 2015;2(6):259–62.

Chapter 13
Roles of Metal Transporters in Cellular Cadmium Transport in Mammals

Seiichiro Himeno and Hitomi Fujishiro

Abstract The mechanisms underlying tissue cadmium (Cd) accumulation in mammals have long been investigated with a focus on the roles of metallothionein, a high-affinity Cd-binding protein. However, the precise mechanisms underlying the influx and efflux of Cd at the cell membrane have remained unclear. As Cd is not an essential element, the transport pathways for essential elements are thought to be utilized for Cd transport, especially for entry into cells. Recently, the roles of metal transporters for iron, calcium, zinc, and manganese in cellular Cd transport have begun to be elucidated. The expression and roles of these transporters vary depending on the tissue or cell types and the nutritional statuses of the elements. Consequently, the usage of metal transporters for Cd transport is also altered depending on the tissue and nutritional status of an element. In this chapter, recent advances in understanding the mechanisms underlying cellular Cd transport from the standpoint of the roles of metal transporters are discussed.

Keywords Cadmium · Transport · Iron · Calcium · Zinc · Manganese · Intestine · Liver · Kidney

13.1 Introduction

In a report published in 1976, Friberg and coworkers demonstrated that the biological half-life of Cd in human kidney is more than 25 years [1]. This was very important because it explained why Cd accumulates preferentially in the kidney after long-term exposure via the diet even though daily doses of Cd are at low levels [1–5].

Especially in Japan, renal Cd concentrations among the general population are much higher than those of non-Asian populations (Fig. 13.1), probably due to the

S. Himeno (✉) · H. Fujishiro
Laboratory of Molecular Nutrition and Toxicology, Faculty of Pharmaceutical Sciences, Tokushima Bunri University, Tokushima, Japan
e-mail: himenos@ph.bunri-u.ac.jp

S. Himeno, K. Aoshima (eds.), *Cadmium Toxicity*, Current Topics in Environmental Health and Preventive Medicine,
https://doi.org/10.1007/978-981-13-3630-0_13

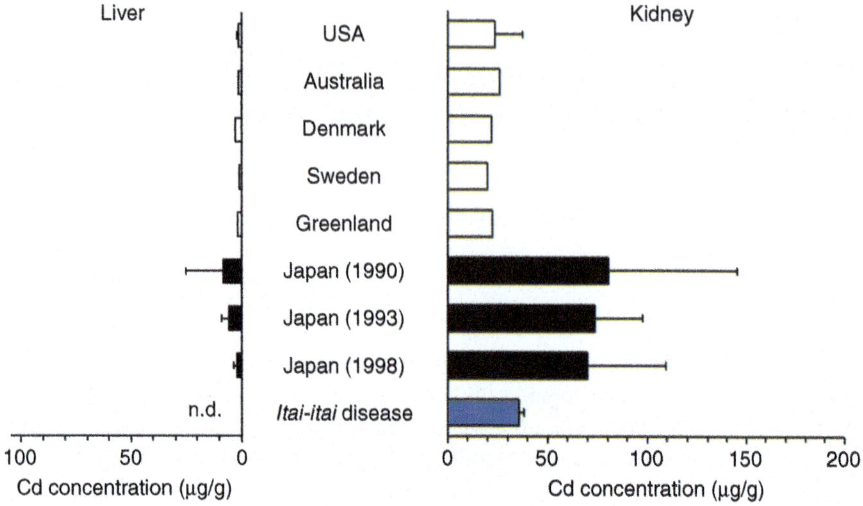

Fig. 13.1 Average Cd concentrations in the liver and kidney in general populations (40–60 years old) in Japan and other countries. The data were obtained from forensic samples collected in the USA [2], Australia [3], Denmark [4], Sweden [1], Greenland [5], and Japan carried out in 1990 [6], 1993 [7], and 1998 [8]. The blue column shows the data of *itai-itai* disease patients [13]. *n.d.* not determined

high lifelong consumption by Japanese of rice and fish, which contain rather high concentrations of Cd [6–8]. The large variation of renal Cd concentrations among the general population in Japan can be partly explained by age-dependent increases in renal Cd concentrations [8]. It has been considered that the critical concentration of Cd in the kidney for causing renal dysfunction is 200 µg Cd/g [9, 10], but some studies have suggested that the safety level of renal Cd should be 50 µg/g calculated from the urinary Cd excretion level [1, 11]. As shown in Fig. 13.1, the average concentrations of renal Cd among general populations of Japan reported in three studies all exceeded 50 µg/g. A portion of individuals showing Cd concentrations in the range of 100–150 µg/g may be nearly at risk of renal dysfunction. Intriguingly, renal Cd concentrations among *itai-itai* disease patients are lower than the average Cd concentrations among the general population (Fig. 13.1) [12, 13], probably because heavily damaged renal cells have already been lost [14].

Thus, the elucidation of the mechanism underlying renal Cd accumulation, which could be a consequence of the influx and efflux of Cd in renal cells, should be the most important issue in the study of Cd toxicity. However, the extraordinarily long biological half-life of Cd itself has hindered studies on the transport mechanism of Cd at the cellular and molecular levels.

Metallothionein (MT) is another factor that has paradoxically hampered the study of Cd transport. MT is a low-molecular-weight protein that plays important roles in the sequestration of Cd and, therefore, in the defense against Cd cytotoxicity [15]. As MT efficiently traps and inactivates Cd within cells due to its high affinity for Cd, the protection against Cd cytotoxicity by MT shows a unique feature: the cellular concentrations of Cd are rather elevated, but the cytotoxicity of Cd is diminished. Thus, it is difficult to study the precise rates of the influx and efflux of Cd in the presence of MT, which quickly traps the Cd ions incorporated into cells.

Because Cd is not an essential element for organisms, there is no specific pathway for Cd to enter cells. Thus far, the roles of the transport systems for other essential elements such as calcium (Ca), iron (Fe), and zinc (Zn) have been investigated as candidates for cellular Cd uptake. Recently, the entities of the transporters for these elements have begun to be elucidated at the molecular level. In this section, the recent advances in understanding the mechanisms underlying cellular Cd transport from the standpoint of the roles of metal transporters are described.

13.2 Involvement of the Pathways for Other Elements in Cadmium Transport

13.2.1 Iron and Iron Transporters

Earlier studies have already demonstrated that experimental animals fed a Fe-deficient diet showed enhanced Cd accumulation [16, 17]. In 1996, Gunshin et al. identified divalent metal transporter 1 (DMT1) as the transporter responsible for divalent Fe (Fe^{2+}) in the intestine [18]. DMT1 also showed affinities for other divalent metals, including Zn^{2+}, Mn^{2+}, and Cd^{2+}. When the animals were subjected to Fe deficiency, the expression of intestinal DMT1 was highly enhanced [19]. This regulation is controlled via the stabilization of DMT1 mRNA by the iron regulatory protein (IRP) and iron-responsive element (IRE). IRP is activated by intracellular Fe deficiency and binds to the IRE located in the 3′-untranslated region of the mRNA of DMT1, leading to the stabilization of the mRNA and, consequently, to the increase in DMT1 protein [20]. When rats were fed an Fe-deficient diet, the level of DMT1 mRNA in the duodenum increased markedly, and the intestinal absorption of Cd was significantly enhanced [21–23]. These findings suggest that DMT1 plays a crucial role in intestinal Cd absorption, especially during times of physiological Fe deficiency, such as pregnancy [24], or of pathological Fe deficiency caused by anemia [22].

The role of the Fe transport system in renal Cd accumulation is different from that in the intestine. In the blood, Fe is bound to the carrier protein transferrin (Tf), which binds the trivalent form of Fe (Fe^{3+}), and the transfer of Fe from the Tf to cells is mediated by endocytosis of the Tf via Tf receptor (TfR) and the subsequent internalization of the Tf-TfR complex into the endosomes of cells [25]. DMT1 in cultured proximal tubule epithelial cells was found to be expressed highly in the endosomes, where the Fe^{3+} released from the internalized Tf-TfR complex is reduced to Fe^{2+} by STEAP1, suggesting that the major role of endosomal DMT1 in renal cells is to transport the Fe^{2+}, which is released from the internalized Tf in the endosomes, into the cytosol [26]. There is a possibility that the Cd^{2+} released from the Cd-MT, which is also internalized into the endosomes of renal cells by endocytosis, is transported to the cytosol via DMT1. In support of this supposition, a study showed that the downregulation of DMT1 by siRNA transfection in cultured renal cells resulted in a decrease in the toxicity of Cd-MT, suggesting that the cytosolic Cd^{2+} transported from the endosomal Cd-MT via DMT1 may partly be involved in renal Cd toxicity [27].

13.2.2 Calcium and Calcium Channels

Since the ionic radius of Cd is identical to that of Ca, it has long been thought that some of the Ca channels may be involved in cellular Cd uptake. However, as the expression of Ca channels is limited to neuronal and other stimuli-responsive cells, the roles of Ca channels in vivo remained unclear. It has been recognized, however, that the Ca-transport system in the intestine is definitely involved in Cd transport since the animals fed a low-Ca diet showed increased Cd accumulation [28, 29]. Min et al. studied the roles of CaT1, also known as TRPV6 (a member of the transient receptor potential family of membrane protein), in intestinal Cd absorption [29, 30]. When mice were fed a Ca-deficient diet, the expression of CaT1 increased up to tenfold that of control mice, and Ca deficiency significantly increased hepatic and renal accumulation of Cd. As CaT1 is highly expressed in the intestine under physiological conditions, these data suggest that CaT1 mediates part of Cd absorption in the intestine.

Several studies using cultured cells indicated the involvement of voltage-dependent Ca channels in Cd uptake. Leslie et al. established a Cd-resistant cell line from mouse embryonic fibroblast (MEF) cells, derived from MT-I/II knockout mice, and found that the expression of $Ca_V3.1$, a component of the T-type Ca channel, is significantly suppressed in the resistant cells [31]. We also established Cd-resistant cells from MT-null MEF cells but failed to find the downregulation of $Ca_V3.1$ [32]. Instead, we found the downregulation of ZIP8 (Zrt, Irt-related protein 8) and ZIP14 in these resistant cells [32, 33]. We also developed Cd-resistant cells from MT-expressing MEF cells and found the downregulation of $Ca_V3.1$ and $Ca_V1.2$, the latter being a component of L-type Ca channel, in addition to that of ZIP8 and DMT1 [34]. The changes in the expression of Ca channels in the MT-expressing Cd-resistant cells, but not in our MT-null Cd-resistant cells, may be related to the higher IC_{50} value of Cd (200 μM) in the MT-expressing Cd-resistant cells than in the MT-null Cd-resistant cells (30 μM) (Table 13.1). It seems likely that during the process of the acquisition of resistance to high concentrations of Cd in the MT-expressing cells, the expression of multiple genes including Ca channels might have been altered in the surviving cells.

13.2.3 ZIP8 and ZIP14 as Cadmium, Manganese, and Zinc Transporters

Interactions between Zn and Cd have been extensively studied in terms of the roles of MT in the deposition of Cd in tissues and cells. Several studies have also indicated that Zn may share the same pathway for entering cells with Cd [35, 36]. We and Prof. Nebert's group at Cincinnati University have identified ZIP8 and ZIP14 as the Zn transporters that are also used for Cd uptake [33, 37, 38].

As described above, we have developed Cd-resistant cell lines from both MT-null and MT-expressing MEF cells and found that the expression of ZIP8 is markedly suppressed in both resistant cells [33, 34]. Prior to the establishment of our Cd-resistant cells from MT-null cells, numerous Cd-resistant cells had been developed, but most studies have attributed Cd resistance to overexpression of the MT gene in resistant cells [39–41]. Utilization of MT-null cells to identify non-MT factors for Cd resistance enabled us to develop a novel type of Cd-resistant cells that acquired Cd resistance due to the reduced accumulation of Cd [42]. This decreased accumulation was found to be caused primarily by the lowered influx of Cd in resistant cells [42, 43]. Because there is no specific pathway for nonessential Cd to enter cells, we assumed that the pathway for the other essential element(s) should be suppressed in the MT-null Cd-resistant cells, resulting in an accompanying suppression of Cd uptake. To identify the element or elements that share this pathway with Cd, we compared the uptake rates of 20 elements simultaneously between Cd-resistant and parental MEF cells by using a multi-tracer technique developed at RIKEN, Japan. We found a marked decrease in the uptake of Mn^{2+} in Cd-resistant MT-null cells at Cd concentrations lower than 10 μM, suggesting the involvement of a high-affinity Mn^{2+} transporter in Cd uptake [43]. A microarray analysis between MT-null Cd-resistant and MT-null parental cells revealed that *Slc39a8* and *Slc39a14* genes coding ZIP8 and ZIP14, respectively, are downregulated markedly in resistant cells [32].

During the time when we were carrying out gene hunting for the Cd/Mn transporter, Nebert and coworkers also identified *Slc39a8* as the gene responsible for the strain difference in the sensitivity to Cd-induced testicular hemorrhage [37]. Kinetic analyses of metal incorporation into the cells, in which ZIP8 or ZIP14 is ectopically expressed, have proved that ZIP8 and ZIP14 are the transporters for Cd^{2+}, Mn^{2+}, and Zn^{2+} [38, 44]. The amino acid sequence of ZIP14 has the highest similarity to ZIP8 among ZIP family members [45], and ZIP14 also has the ability to transport Cd^{2+}, Mn^{2+}, and Zn^{2+} [38]. ZIP14 has another characteristic: it is efficiently inducible by interleukin-6 (IL-6), and the induced ZIP14 in the liver was shown to be responsible for the enhanced uptake of serum Zn, resulting in lowered levels of serum Zn at the time of inflammation [46].

The roles of ZIP8 in the cellular uptake of both Cd ad Mn have been clearly shown in rat basophilic leukemia RBL-2H3 cells, which exhibited high expression of ZIP8 and high sensitivity to Cd and Mn toxicity due to their efficient accumulation of both metals [47]. The downregulation of ZIP8 by siRNA transfection resulted

Table 13.1 Expression profiles of metal transporters in Cd-resistant cell lines

Cells	IC_{50} of Cd	MT	ZIP8	ZIP14	DMT1
MT−/− Cd-r	30 μM	Null	⇩	⇩	⇩
MT+/+ Cd-r	200 μM	⇧	⇩	–	⇩
RBL-Cdr	350 μM	⇧	⇩	–	–
RBL-Mnr	77 μM	–	⇩	⇩	–

in both decreased accumulation and lower toxicity of both metals. To further explore the roles of ZIP8 in the uptake of Cd and Mn in RBL-2H3 cells, we developed Cd-resistant and Mn-resistant cells from RBL-2H3 cells [48]. Both the Cd-resistant and Mn-resistant RBL-2H3 cells downregulated ZIP8 but not ZIP14 or DMT1. Cd-resistant RBL-2H3 cells showed a cross-resistance to Mn due to the reduced accumulation of both Cd and Mn, and Mn-resistant RBL-2H3 cells showed a cross-resistance to Cd due similarly to the reduced accumulation of Mn and Cd (Table 13.1). These results suggest that ZIP8 plays a greater role than ZIP14 in the uptake of Cd and Mn, at least in cultured cells.

13.3 Tissue-Specific Roles of Metal Transporters in the Transport of Cadmium

13.3.1 Intestinal Transport of Cadmium

Many in vivo studies on the effects of dietary Fe or Ca deficiency have demonstrated that the transport systems for both Fe and Ca are involved in intestinal Cd absorption [21–23, 28–34]. DMT1 and CaT1 are highly expressed in the apical membrane of the enterocytes in humans and experimental animals, and their expression can be induced by dietary deficiencies of Fe and Ca, respectively. Thus, these transporters may play significant roles in the apical uptake of Cd from the intestinal tract (Fig. 13.2). However, the mechanisms by which Cd incorporated into enterocytes is transferred to the blood remain unsolved.

Ferroportin is a transporter responsible for the efflux of Fe^{2+} from the enterocytes to the blood [49]. At the basal membrane of the enterocytes, ferroportin is expressed in conjunction with hephastin, which can oxidize Fe^{2+} to Fe^{3+}, so that the excreted Fe^{2+} is readily transferred to the circulating Tf, which binds only Fe^{3+}. It remains unknown whether this system of exporting Fe is used also for the transfer of Cd into the blood.

In the case of Ca, the absorbed Ca in the enterocytes is known to be bound to calbindin, a Ca-binding protein, and then excreted into the blood [50]. However, no

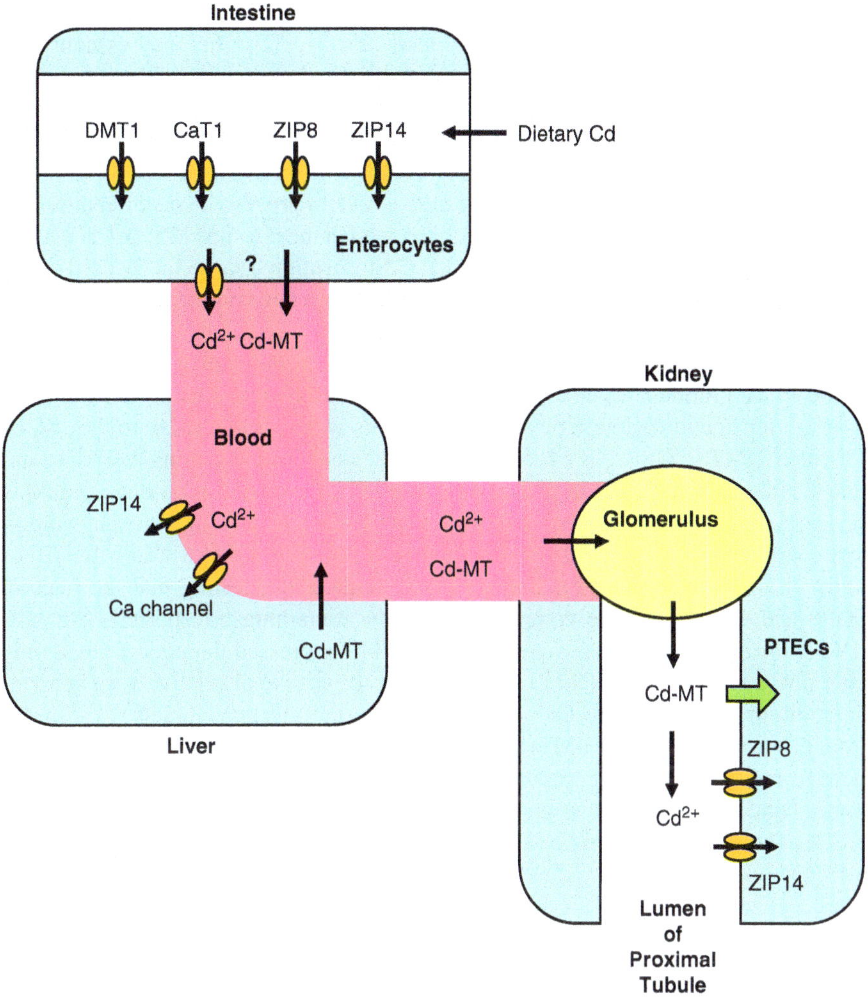

Fig. 13.2 Candidate metal transporters involved in the transport of Cd in the intestine, liver, and kidney. In the intestine, dietary Cd is absorbed into the enterocytes via either DMT1, CaT1, ZIP8, or ZIP14 depending on the nutritional status of Fe, Ca, Zn, and Mn. The pathways of Cd and Cd-MT from enterocytes to blood remain unknown. In the liver, ZIP14 and Ca channels may be involved in Cd^{2+} absorption, and the contribution of endocytosis via sinusoidal epithelial cells may be minimal. Cd^{2+} in the blood may be bound to low-molecular-weight biomolecules. In the kidney, Cd-MT can be filtered through the glomerulus and incorporated into PTECs (proximal tubule epithelial cells) via endocytosis (green arrow). A portion of Cd existing as Cd^{2+} could be incorporated into PTECs via ZIP8 or ZIP14

information is available about whether Cd can be bound to calbindin and exported to the blood in a manner similar to that of Ca. It seems unlikely, however, that the transfer of Cd into the blood from the enterocytes is mediated by the efficient transport systems for Fe and Ca, because the efficiency of intestinal Cd absorption is known to be as low as 5%.

Many studies have shown that Cd induces MT in the intestine and suggested that Cd bound to MT is transferred to the blood [29, 51, 52]. This may explain why dietary Cd is preferentially deposited in the kidney compared with parenterally injected $CdCl_2$, because Cd-MT is efficiently incorporated into the proximal tubule epithelial cells via endocytosis [53].

Dietary Zn deficiency enhances the efficiency of intestinal Cd uptake [54, 55], suggesting that Zn transporters such as ZIP8 or ZIP14 are also involved in intestinal Cd uptake. ZIP14 is known to be highly expressed in the intestine [35, 56]. However, most studies on the roles of Zn in intestinal Cd absorption have so far focused on the roles of Zn in MT induction, and little information has been provided on the roles of Zn transporters in the intestinal transport of Cd.

To explore the roles of these metal transporters in the uptake of Cd under the same concentrations, we examined the effects of siRNA transfection of each metal transporter in cultured enterocytes. Caco-2 cells were transfected with the siRNA of either DMT1, ZIP8, ZIP14, or CaT1 and then cultured in a trans-well to examine the rates of Cd uptake from the apical side [57]. Low concentrations of Cd (0.1, 1, and 5 μM) were added to the medium of the apical side to mimic the physiological conditions of dietary Cd uptake. Downregulation of ZIP8 or CaT1 by siRNA did not alter the uptake of Cd at any concentration, suggesting that the roles of ZIP8 and CaT1 could be compensated by the remaining transporter(s) such as DMT1 or ZIP14. The downregulation of DMT1 suppressed the uptake of Cd only at 5 μM, suggesting that DMT1 is involved in the uptake of relatively higher concentrations of Cd. As the Cd concentration in the intestinal lumen is estimated roughly to be less than 1 μM when the diet contains 0.1 ppm Cd [57], the role of DMT1 may become more important when the diet is contaminated by Cd. On the other hand, the downregulation of ZIP14 suppressed the uptake of Cd at 0.1 and 1 μM. These data may implicate the role of ZIP14 at lower dietary concentrations of Cd, but more detailed studies are required because there is a report showing that the localization of ZIP14 in the intestinal cells is limited to the basal side of the enterocytes [56].

13.3.2 Cd Transport in the Liver

Compared to the intestine and kidney, fewer studies have been carried out regarding Cd transport systems in the liver. Several studies have indicated the involvement of Ca channels [58, 59] and ZIP14 in the uptake of Cd into hepatic cells [60] (Fig. 13.2). Hepatic ZIP14 is known to be induced by inflammatory stimuli such as IL-6, resulting in an enhanced uptake of serum Zn into the liver [46]. Similarly, several types of inflammatory stimuli were shown to enhance the accumulation of Cd in the mouse liver via the induction of ZIP14 [61]. In ZIP14 knockout mice, the accumulation of orally administered Cd in the liver was decreased to about half of that in the control mice, suggesting that ZIP14 may play a significant role in the uptake of Cd into the liver even under conditions of no inflammation [60].

After the ingestion of dietary Cd, the site of Cd uptake into the parenchymal cells of liver is the sinusoid. A portion of the Cd in blood in the portal vein may exist as an ionic form of Cd^{2+}, which can be incorporated into the liver via Zn transporters [60, 62, 63] or Ca channels [58, 59]. However, many studies have suggested that the Cd bound to MT also exists in the portal vein blood because MT is highly induced in the intestine by the ingestion of dietary Cd [29, 51, 52]. Although there exists an endocytosis system of proteins in the sinusoidal endothelial cells of the liver [64], it seems unlikely that Cd-MT is efficiently incorporated into hepatocytes via endocytosis because the direct injection of Cd-MT into the blood has been shown to result in preferential deposition in the kidney but not in the liver [65, 66]. Presumably, the Cd-MT that evaded endocytosis in the sinusoid of the liver circulates in the blood and then reaches the kidney, where efficient endocytic incorporation of Cd-MT into the proximal tubule cells is observed [53]. In support of this notion, when Cd-MT is directly added to the ligated intestinal loop in rat, the deposition of Cd was preferentially detected in the kidney, whereas the addition of $CdCl_2$ resulted in higher deposition of Cd in the liver [67].

When Cd accumulates in the liver, the production of MT protein is highly induced, resulting in a longer retention of Cd in the liver. If the deposited Cd begins to cause liver damage, Cd-MT in hepatic cells is released into the blood in a similar way as the release of hepatocellular transaminases from damaged hepatic cells [68]. Under physiological conditions, the release rate of Cd-MT from hepatic cells into the blood is thought to be very low. A portion of the Cd is excreted into the bile as glutathione conjugates via the canalicular membranes of hepatic cells, although the contribution of biliary Cd excretion to the reduction of hepatic Cd retention was shown to be minimal [69].

13.3.3 Cd Transport in the Kidney

Mounting evidence has demonstrated that MT is the key player in renal Cd deposition [53, 70]. MT carrying Cd in blood plasma can be filtered through the glomerulus due to its small size (−7000 kDa) and is reabsorbed into the proximal tubule epithelial cells from the lumen via endocytosis. Direct injection of purified Cd-MT into the blood has been shown to cause preferential accumulation of Cd in the kidney [65, 66]. The endocytosis of Cd-MT is mediated by megalin, a receptor for a wide variety of proteins expressed at the apical membrane of proximal tubule epithelial cells [71–73]. However, pathways of Cd incorporation into renal cells other than endocytosis of Cd-MT may also be functioning in the kidney [74] (Fig. 13.2).

Since the proximal tubule epithelial cells are polarized cells having both apical and basolateral transport systems, we examined the transport of Cd in immortalized mouse proximal tubule (PT) cells by using a trans-well culture system, in which the influx and efflux of the metals via apical and basolateral membranes of cells can be measured separately [75]. The results showed that the uptake rate of Cd from the apical side was much higher than that from the basolateral side of PT cells. The excretion of incorporated Cd was detected almost exclusively on the apical side of the cells, though the efficiency of excretion was low.

To test the role of each metal transporter in the uptake of Cd, siRNAs of ZIP8, ZIP14, or DMT1 were transfected into PT cells, and then the cells were cultured on the trans-well [75]. After the confirmation of the tight junction formation, the efficiencies of Cd uptake from the apical side of the cells were examined. The down-regulation of ZIP8 and ZIP14 resulted in decreases in the uptake rates of Cd by about 50% vs. the control cells. The addition of Mn^{2+} or Zn^{2+} in the medium of the apical side competitively suppressed the uptake of Cd^{2+}. These results suggest that ZIP8 and ZIP14 play significant roles in the uptake of Cd from the apical side of the proximal tubule cells. The transfection of DMT1 siRNA also reduced the apical uptake of Cd but to a lesser extent than those of ZIP8 and ZIP14. As described previously, DMT1 in renal cells may be involved in intracellular translocation of Cd rather than in the uptake of Cd from the lumen or blood [27].

The expression of metal transporters in the kidney was also examined in an in vivo system. In situ hybridization of ZIP8 and ZIP14 detected the mRNAs of ZIP8 and ZIP14 highly in the border region of the cortex and medulla, with concomitant PAS staining indicative of the proximal tubules. The proximal tubules in this region are composed mainly of the S3 segment [75]. The endocytosis of Cd-MT and the consequent cytotoxicity are known to occur mainly in the convoluted proximal tubules, which are composed of the S1 and S2 segments [76]. According to these in vitro and in vivo observations, we postulated the hypothesis that ZIP8 and ZIP14 located in the S3 segment of the proximal tubules play a significant role in the reabsorption of the ionic form of Cd in the lumen. The sources of luminal Cd^{2+} may include Cd bound loosely to low-molecular-weight substances such as citrate or bicarbonate (the latter was shown to enhance the efficiency of ZIP8-mediated Cd uptake [47]) as well as Cd^{2+} released from the S1 and S2 segments of proximal tubule cells, as observed in our experiment with the trans-well culture of the PT cells. Figure 13.3 shows a schematic of our hypothesis. To prove this hypothesis, experiments with immortalized S1, S2, and S3 cells are under way in our laboratory.

We also observed that siRNA transfection of ZIP8 or ZIP14 significantly suppressed the uptake of Mn^{2+} in PT cells [75]. Recent studies on mutations in human ZIP8 have highlighted ZIP8's role in the reabsorption of Mn in the kidney. Several cases of patients having mutations in *SLC39A8* coding ZIP8 were found in Germany and Egypt [77, 78]. Blood Mn concentrations in the patients were reduced to undetectable levels. As Mn plays a crucial role in the enzyme activity of galactosyl transferase, which is essential for the synthesis of sugar chains in all proteins, patients with a ZIP8 mutation exhibited a syndrome similar to that of inherited glycosylation defects. The supplementation of Mn to such patients was effective, but higher levels of excretion of Mn were observed in their urine [79], suggesting that ZIP8 has a significant role in the reabsorption of Mn from the lumen into the kidney.

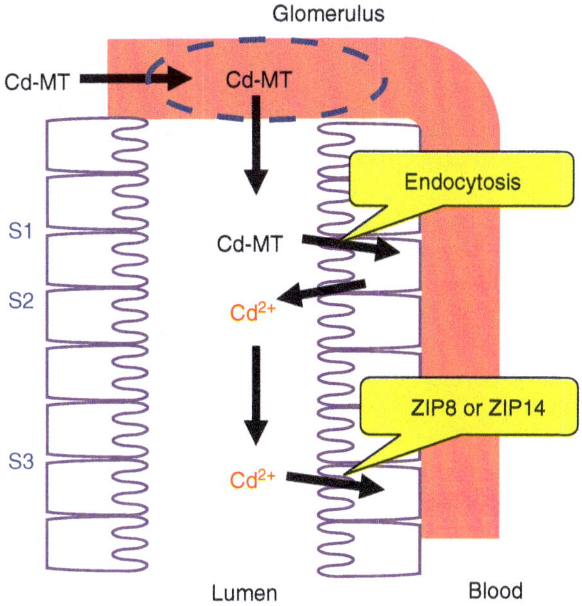

Fig. 13.3 Hypothesis of dynamic renal handling of Cd. The Cd-MT filtered through the glomerulus is reabsorbed via megalin-dependent endocytosis into PTECs in the S1 and S2 segments. The filtered or secreted Cd^{2+} in the lumen could be incorporated into PTECs in the S3 segment, where ZIP8 and ZIP14 are highly expressed

13.4 Conclusion

Previously, most studies on tissue Cd deposition have been undertaken from the standpoint of MT's functions and roles. However, recent advances in understanding the detailed functions and regulation mechanisms of metal transporters have enabled us to recognize more important roles of DMT1, ZIP8, and ZIP14 than before in the deposition and handling of Cd in the organs, especially the intestine and kidney. Recently, several studies in large populations of humans have reported that polymorphisms of ZIP8 are associated with alterations in blood metal levels and with disease occurrence [80–84]. Further studies are warranted to elucidate the effects of genetic polymorphisms of DMT1, ZIP8, and ZIP14 on the efficiency of Cd transport and the resultant health effects. In addition, further studies are required to clarify the roles of metal transporters located in subcellular organelle to understand the translocation mechanisms of Cd within cells and the relationship between the translocation and cytotoxicity of Cd.

References

1. Elinder CG, Lind B, Kjellström T, Linnman L, Friberg L. Cadmium in kidney cortex, liver, and pancreas from Swedish autopsies. Estimation of biological half time in kidney cortex, considering calorie intake and smoking habits. Arch Environ Health. 1976;31:292–302.
2. Kjellström T. Exposure and accumulation of cadmium in populations from Japan, the United States, and Sweden. Environ Health Perspect. 1979;28:169–97.
3. Satarug S, Baker JR, Urbenjapol S, Haswell-Elkins M, Reilly PE, Williams DJ, Moore MR. A global perspective on cadmium pollution and toxicity in non-occupationally exposed population. Toxicol Lett. 2003;137:65–83.
4. Hansen JC, Gron P, Jespersen BA, Voigt J, Simonsen J, Dalgaard JB, Hansen E. Cadmium exposure in Denmark. Based on analyses of liver and kidney tissues. Dan Med Bull. 1989;36:499–502.
5. Johansen P, Mulvad G, Pedersen HS, Hansen JC, Riget F. Accumulation of cadmium in livers and kidneys in Greenlanders. Sci Total Environ. 2006;372:58–63.
6. Yoshinaga J, Matsuo N, Imai H, Nakazawa M, Suzuki T, Morita M, Akagi H. Interrelationship between the concentrations of some elements in the organs of Japanese with special reference to selenium-heavy metal relationships. Sci Total Environ. 1990;91:127–40.
7. Noda H, Sugiyama S, Yamaguchi M, Tatsumi S, Sano Y, Konishi S, Furutani A, Yoshimura M. Study on secular changes of cadmium concentration accumulated in main organs of Japanese. Nihon Hoigaku Zasshi. 1993;47:153–9.
8. Yoshida M, Ohta H, Yamauchi Y, Seki Y, Sagi M, Yamazaki K, Sumi Y. Age-dependent changes in metallothionein levels in liver and kidney of the Japanese. Biol Trace Elem Res. 1998;63:167–75.
9. Kjellström T, Elinder CG, Friberg L. Conceptual problems in establishing the critical concentration of cadmium in human kidney cortex. Environ Res. 1984;33:284–95.
10. Roels H, Lauwerys R, Dardenne AN. The critical level of cadmium in human renal cortex: a reevaluation. Toxicol Lett. 1983;15:357–60.
11. Buchet JP, Lauwerys R, Roels H, Bernard A, Bruaux P, Claeys F, Ducoffre G, de Plaen P, Staessen J, Amery A, Lijnen P, Thijs L, Rondia D, Sartor F, Saint Remy A, Nick L. Renal effects of cadmium body burden of the general population. Lancet. 1990;336:699–702.
12. Hayashi C, Koizumi N, Nishio H, Ikeda M. Cadmium and other metal levels in autopsy samples from a cadmium-polluted area and non-polluted control areas in Japan. Biol Trace Elem Res. 2012;145:10–22.
13. Baba H, Tsuneyama K, Kumada T, Aoshima K, Imura J. Histopathological analysis for osteomalacia and tubulopathy in itai-itai disease. J Toxicol Sci. 2014;39:91–6.
14. Yasuda M, Miwa A, Kitagawa M. Morphometric studies of renal lesions in Itai-itai disease: chronic cadmium nephropathy. Nephron. 1995;69:14–9.
15. Klaassen CD, Liu J, Diwan BA. Metallothionein protection of cadmium toxicity. Toxicol Appl Pharmacol. 2009;238:215–20.
16. Ragan HA. Effects of iron deficiency on the absorption and distribution of lead and cadmium in rats. J Lab Clin Med. 1977;90:700–6.
17. Flanagan PR, McLellan JS, Haist J, Cherian G, Chamberlain MJ, Valberg LS. Increased dietary cadmium absorption in mice and human subjects with iron deficiency. Gastroenterology. 1978;74:841–6.
18. Gunshin H, Mackenzie B, Berger UV, Gunshin Y, Romero MF, Boron WF, Nussberger S, Gollan JL, Hediger MA. Cloning and characterization of a mammalian proton-coupled metal-ion transporter. Nature. 1997;388:482–8.
19. Morgan EH, Oates PS. Mechanisms and regulation of intestinal iron absorption. Blood Cells Mol Dis. 2002;29:384–99.
20. Theil EC. Iron homeostasis and nutritional iron deficiency. J Nutr. 2011;141:724S–8S.
21. Park JD, Cherrington NJ, Klaassen CD. Intestinal absorption of cadmium is associated with divalent metal transporter 1 in rats. Toxicol Sci. 2002;68:288–94.

22. Min KS, Iwata N, Tetsutikawahara N, Onosaka S, Tanaka K. Effect of hemolytic and iron-deficiency anemia on intestinal absorption and tissue accumulation of cadmium. Toxicol Lett. 2008;179:48–52.
23. Min KS, Ueda H, Kihara T, Tanaka K. Increased hepatic accumulation of ingested Cd is associated with upregulation of several intestinal transporters in mice fed diets deficient in essential metals. Toxicol Sci. 2008;106:284–9.
24. Leazer TM, Liu Y, Klaassen CD. Cadmium absorption and its relationship to divalent metal transporter-1 in the pregnant rat. Toxicol Appl Pharmacol. 2002;185:18–24.
25. Dautry-Varsat A. Receptor-mediated endocytosis: the intracellular journey of transferrin and its receptor. Biochimie. 1986;68:375–81.
26. Abouhamed M, Gburek J, Liu W, Torchalski B, Wilhelm A, Wolff NA, Christensen EI, Thévenod F, Smith CP. Divalent metal transporter 1 in the kidney proximal tubule is expressed in late endosomes/lysosomal membranes: implications for renal handling of protein-metal complexes. Am J Physiol Renal Physiol. 2006;290:F1525–33.
27. Abouhamed M, Wolff NA, Lee WK, Smith CP, Thévenod F. Knockdown of endosomal/lysosomal divalent metal transporter 1 by RNA interference prevents cadmium-metallothionein-1 cytotoxicity in renal proximal tubule cells. Am J Physiol Renal Physiol. 2007;293:F705–12.
28. Washko PW, Cousins RJ. Metabolism of ^{109}Cd in rats fed normal and low-calcium diets. J Toxicol Environ Health. 1976;1:1055–66.
29. Min KS, Ueda H, Tanaka K. Involvement of intestinal calcium transporter 1 and metallothionein in cadmium accumulation in the liver and kidney of mice fed a low-calcium diet. Toxicol Lett. 2008;176:85–92.
30. Min KS, Sano E, Ueda H, Sakazaki F, Yamada K, Takano M, Tanaka K. Dietary deficiency of calcium and/or iron, an age-related risk factor for renal accumulation of cadmium in mice. Biol Pharm Bull. 2015;38:1557–63.
31. Leslie EM, Liu J, Klaassen CD, Waalkes MP. Acquired cadmium resistance in metallothionein-I/II(−/−) knockout cells: role of the T-type calcium channel Cacnα_{1G} in cadmium uptake. Mol Pharmacol. 2006;69:629–39.
32. Fujishiro H, Okugaki S, Nagao S, Satoh M, Himeno S. Characterization of gene expression profiles of metallothionein null cadmium-resistant cells. J Health Sci. 2006;52:292–9.
33. Fujishiro H, Okugaki S, Kubota K, Fujiyama T, Miyataka H, Himeno S. The role of ZIP8 down-regulation in cadmium-resistant metallothionein-null cells. J Appl Toxicol. 2009;29:367–73.
34. Fujishiro H, Kubota K, Inoue D, Inoue A, Yanagiya T, Enomoto S, Himeno S. Cross-resistance of cadmium-resistant cells to manganese is associated with reduced accumulation of both cadmium and manganese. Toxicology. 2011;280:118–25.
35. Waalkes MP, Perantoni A. In vitro assessment of target cell specificity in cadmium carcinogenesis: interactions of cadmium and zinc with isolated interstitial cells of the rat testes. In Vitro Cell Dev Biol. 1988;24:558–65.
36. Barbier O, Dauby A, Jacquillet G, Tauc M, Poujeol P, Cougnon M. Zinc and cadmium interactions in a renal cell line derived from rabbit proximal tubule. Nephron Physiol. 2005;99:p74–84.
37. Dalton TP, He L, Wang B, Miller ML, Jin L, Stringer KF, Chang X, Baxter CS, Nebert DW. Identification of mouse SLC39A8 as the transporter responsible for cadmium-induced toxicity in the testis. Proc Natl Acad Sci U S A. 2005;102:3401–6.
38. Girijashanker K, He L, Soleimani M, Reed JM, Li H, Liu Z, Wang B, Dalton TP, Nebert DW. *Slc39a14* gene encodes ZIP14, a metal/bicarbonate symporter: similarities to the ZIP8 transporter. Mol Pharmacol. 2008;73:1413–23.
39. Grady DL, Moyzis RK, Hildebrand CE. Molecular and cellular mechanisms of cadmium resistance in cultured cells. Experientia Suppl. 1987;52:447–56.
40. Li W, Kagan HM, Chou IN. Alterations in cytoskeletal organization and homeostasis of cellular thiols in cadmium-resistant cells. Toxicol Appl Pharmacol. 1994;126:114–23.
41. Crawford BD, Enger MD, Griffith BB, Griffith JK, Hanners JL, Longmire JL, Munk AC, Stallings RL, Tesmer JG, Walters RA, et al. Coordinate amplification of metallothionein I and II genes in cadmium-resistant Chinese hamster cells: implications for mechanisms regulating metallothionein gene expression. Mol Cell Biol. 1985;5:320–9.

42. Yanagiya T, Imura N, Kondo Y, Himeno S. Reduced uptake and enhanced release of cadmium in cadmium-resistant metallothionein null fibroblasts. Life Sci. 1999;65:PL177–82.
43. Yanagiya T, Imura N, Enomoto S, Kondo Y, Himeno S. Suppression of a high-affinity transport system for manganese in cadmium-resistant metallothionein-null cells. J Pharmacol Exp Ther. 2000;292:1080–6.
44. He L, Girijashanker K, Dalton TP, Reed J, Li H, Soleimani M, Nebert DW. ZIP8, member of the solute-carrier-39 (SLC39) metal-transporter family: characterization of transporter properties. Mol Pharmacol. 2006;70:171–80.
45. Kambe T, Tsuji T, Hashimoto A, Itsumura N. The physiological, biochemical, and molecular roles of zinc transporters in zinc homeostasis and metabolism. Physiol Rev. 2015;95:749–84.
46. Liuzzi JP, Lichten LA, Rivera S, Blanchard RK, Aydemir TB, Knutson MD, Ganz T, Cousins RJ. Interleukin-6 regulates the zinc transporter Zip14 in liver and contributes to the hypozincemia of the acute-phase response. Proc Natl Acad Sci U S A. 2005;102:6843–8.
47. Fujishiro H, Doi M, Enomoto S, Himeno S. High sensitivity of RBL-2H3 cells to cadmium and manganese: an implication of the role of ZIP8. Metallomics. 2011;3:710–8.
48. Fujishiro H, Ohashi T, Takuma M, Himeno S. Suppression of ZIP8 expression is a common feature of cadmium-resistant and manganese-resistant RBL-2H3 cells. Metallomics. 2013;5:437–44.
49. Drakesmith H, Nemeth E, Ganz T. Ironing out Ferroportin. Cell Metab. 2015;22:777–87.
50. Choi KC, Jeung EB. Molecular mechanism of regulation of the calcium-binding protein calbindin-D9k, and its physiological role(s) in mammals: a review of current research. J Cell Mol Med. 2008;12:409–20.
51. Min KS, Fujita Y, Onosaka S, Tanaka K. Role of intestinal metallothionein in absorption and distribution of orally administered cadmium. Toxicol Appl Pharmacol. 1991;109:7–16.
52. Min KS, Nakatsubo T, Kawamura S, Fujita Y, Onosaka S, Tanaka K. Effects of mucosal metallothionein in small intestine on tissue distribution of cadmium after oral administration of cadmium compounds. Toxicol Appl Pharmacol. 1992;113:306–10.
53. Sabolic I, Breljak D, Skarica M, Herak-Kramberger CM. Role of metallothionein in cadmium traffic and toxicity in kidneys and other mammalian organs. Biometals. 2010;23:897–926.
54. Waalkes MP. Effect of dietary zinc deficiency on the accumulation of cadmium and metallothionein in selected tissues of the rat. J Toxicol Environ Health. 1986;18:301–13.
55. Reeves PG, Chaney RL. Marginal nutritional status of zinc, iron, and calcium increases cadmium retention in the duodenum and other organs of rats fed rice-based diets. Environ Res. 2004;96:311–22.
56. Guthrie GJ, Aydemir TB, Troche C, Martin AB, Chang SM, Cousins RJ. Influence of ZIP14 (slc39A14) on intestinal zinc processing and barrier function. Am J Physiol Gastrointest Liver Physiol. 2015;308:G171–8.
57. Fujishiro H, Hamao S, Tanaka R, Kambe T, Himeno S. Concentration-dependent roles of DMT1 and ZIP14 in cadmium absorption in Caco-2 cells. J Toxicol Sci. 2017;42:559–67.
58. Souza V, Bucio L, Gutierrez-Ruiz MC. Cadmium uptake by a human hepatic cell line (WRL-68 cells). Toxicology. 1997;120:215–20.
59. Souza V, Bucio L, Jay D, Chavez E, Gutierrez-Ruiz MC. Effect of cadmium on calcium transport in a human fetal hepatic cell line (WRL-68 cells). Toxicology. 1996;112:97–104.
60. Jorge-Nebert LF, Galvez-Peralta M, Landero Figueroa J, Somarathna M, Hojyo S, Fukada T, Nebert DW. Comparing gene expression during cadmium uptake and distribution: untreated versus oral Cd-treated wild-type and ZIP14 knockout mice. Toxicol Sci. 2015;143:26–35.
61. Min KS, Takano M, Amako K, Ueda H, Tanaka K. Involvement of the essential metal transporter Zip14 in hepatic Cd accumulation during inflammation. Toxicol Lett. 2013;218:91–6.
62. Shaikh ZA, Blazka ME, Endo T. Metal transport in cells: cadmium uptake by rat hepatocytes and renal cortical epithelial cells. Environ Health Perspect. 1995;103(Suppl 1):73–5.
63. Shimada H, Yasutake A, Hirashima T, Takamure Y, Kitano T, Waalkes MP, Imamura Y. Strain difference of cadmium accumulation by liver slices of inbred Wistar-Imamichi and Fischer 344 rats. Toxicol In Vitro. 2008;22:338–43.

64. Poisson J, Lemoinne S, Boulanger C, Durand F, Moreau R, Valla D, Rautou PE. Liver sinusoidal endothelial cells: Physiology and role in liver diseases. J Hepatol. 2017;66:212–27.
65. Min KS, Onosaka S, Tanaka K. Renal accumulation of cadmium and nephropathy following long-term administration of cadmium-metallothionein. Toxicol Appl Pharmacol. 1996;141:102–9.
66. Dorian C, Gattone VH 2nd, Klaassen CD. Discrepancy between the nephrotoxic potencies of cadmium-metallothionein and cadmium chloride and the renal concentration of cadmium in the proximal convoluted tubules. Toxicol Appl Pharmacol. 1995;130:161–8.
67. Ohta H, Cherian MG. Gastrointestinal absorption of cadmium and metallothionein. Toxicol Appl Pharmacol. 1991;107:63–72.
68. Dudley RE, Gammal LM, Klaassen CD. Cadmium-induced hepatic and renal injury in chronically exposed rats: likely role of hepatic cadmium-metallothionein in nephrotoxicity. Toxicol Appl Pharmacol. 1985;77:414–26.
69. Sugawara N, Lai YR, Arizono K, Kitajima T, Inoue H. Lack of biliary excretion of Cd linked to an inherent defect of the canalicular isoform of multidrug resistance protein (cMrp) does not abnormally stimulate accumulation of Cd in the Eisai hyperbilirubinemic (EHB) rat liver. Arch Toxicol. 1997;71:336–9.
70. Nordberg M, Jin T, Nordberg GF. Cadmium, metallothionein and renal tubular toxicity. IARC Sci Publ. 1992;118:293–7.
71. Klassen RB, Crenshaw K, Kozyraki R, Verroust PJ, Tio L, Atrian S, Allen PL, Hammond TG. Megalin mediates renal uptake of heavy metal metallothionein complexes. Am J Physiol Renal Physiol. 2004;287:F393–403.
72. Wolff NA, Abouhamed M, Verroust PJ, Thévenod F. Megalin-dependent internalization of cadmium-metallothionein and cytotoxicity in cultured renal proximal tubule cells. J Pharmacol Exp Ther. 2006;318:782–91.
73. Onodera A, Tani M, Michigami T, Yamagata M, Min KS, Tanaka K, Nakanishi T, Kimura T, Itoh N. Role of megalin and the soluble form of its ligand RAP in Cd-metallothionein endocytosis and Cd-metallothionein-induced nephrotoxicity *in vivo*. Toxicol Lett. 2012;212:91–6.
74. Thévenod F. Catch me if you can! Novel aspects of cadmium transport in mammalian cells. Biometals. 2010;23:857–75.
75. Fujishiro H, Yano Y, Takada Y, Tanihara M, Himeno S. Roles of ZIP8, ZIP14, and DMT1 in transport of cadmium and manganese in mouse kidney proximal tubule cells. Metallomics. 2012;4:700–8.
76. Dorian C, Gattone VH 2nd, Klaasen CD. Renal cadmium deposition and injury as a result of accumulation of cadmium-metallothionein (CdMT) by the proximal convoluted tubules--A light microscopic autoradiography study with ^{109}CdMT. Toxicol Appl Pharmacol. 1992;114:173–81.
77. Park JH, Hogrebe M, Gruneberg M, DuChesne I, von der Heiden AL, Reunert J, Schlingmann KP, Boycott KM, Beaulieu CL, Mhanni AA, Innes AM, Hortnagel K, Biskup S, Gleixner EM, Kurlemann G, Fiedler B, Omran H, Rutsch F, Wada Y, Tsiakas K, Santer R, Nebert DW, Rust S, Marquardt T. SLC39A8 deficiency: a disorder of manganese transport and glycosylation. Am J Hum Genet. 2015;97:894–903.
78. Boycott KM, Beaulieu CL, Kernohan KD, Gebril OH, Mhanni A, Chudley AE, Redl D, Qin W, Hampson S, Kury S, Tetreault M, Puffenberger EG, Scott JN, Bezieau S, Reis A, Uebe S, Schumacher J, Hegele RA, McLeod DR, Galvez-Peralta M, Majewski J, Ramaekers VT, Nebert DW, Innes AM, Parboosingh JS, Abou Jamra R. Autosomal-recessive intellectual disability with cerebellar atrophy syndrome caused by mutation of the manganese and zinc transporter gene SLC39A8. Am J Hum Genet. 2015;97:886–93.
79. Park JH, Hogrebe M, Fobker M, Brackmann R, Fiedler B, Reunert J, Rust S, Tsiakas K, Santer R, Gruneberg M, Marquardt T. SLC39A8 deficiency: biochemical correction and major clinical improvement by manganese therapy. Genet Med. 2018;20:259–68.
80. Rentschler G, Kippler M, Axmon A, Raqib R, Skerfving S, Vahter M, Broberg K. Cadmium concentrations in human blood and urine are associated with polymorphisms in zinc transporter genes. Metallomics. 2014;6:885–91.

81. Ng E, Lind PM, Lindgren C, Ingelsson E, Mahajan A, Morris A, Lind L. Genome-wide association study of toxic metals and trace elements reveals novel associations. Hum Mol Genet. 2015;24:4739–45.
82. Waterworth DM, Ricketts SL, Song K, Chen L, Zhao JH, Ripatti S, Aulchenko YS, Zhang W, Yuan X, Lim N, Luan J, Ashford S, Wheeler E, Young EH, Hadley D, Thompson JR, Braund PS, Johnson T, Struchalin M, Surakka I, Luben R, Khaw KT, Rodwell SA, Loos RJ, Boekholdt SM, Inouye M, Deloukas P, Elliott P, Schlessinger D, Sanna S, Scuteri A, Jackson A, Mohlke KL, Tuomilehto J, Roberts R, Stewart A, Kesaniemi YA, Mahley RW, Grundy SM, McArdle W, Cardon L, Waeber G, Vollenweider P, Chambers JC, Boehnke M, Abecasis GR, Salomaa V, Jarvelin MR, Ruokonen A, Barroso I, Epstein SE, Hakonarson HH, Rader DJ, Reilly MP, Witteman JC, Hall AS, Samani NJ, Strachan DP, Barter P, van Duijn CM, Kooner JS, Peltonen L, Wareham NJ, McPherson R, Mooser V, Sandhu MS. Genetic variants influencing circulating lipid levels and risk of coronary artery disease. Arterioscler Thromb Vasc Biol. 2010;30:2264–76.
83. Zhang R, Witkowska K, Ng F, Caulfield MJ, Ye S. Lb03.08: Hypertension Related Variant of Solute Carrier Family 39 Member 8 Gene Influences Cadmium Uptake and Cell Toxicity. J Hypertens. 2015;33(Suppl 1):e128.
84. Costas J. The highly pleiotropic gene *SLC39A8* as an opportunity to gain insight into the molecular pathogenesis of schizophrenia. Am J Med Genet B Neuropsychiatr Genet. 2018;177:274–83.

Chapter 14
Disturbance in Phosphorus Metabolism by Cadmium Exposure

Shinsuke Kido, Ichiro Kaneko, and Ken-ichi Miyamoto

Abstract *Itai-itai* disease is an endemic disease characterized by osteomalacia accompanied with osteoporosis and multiple proximal tubular dysfunctions (Fanconi syndrome). *Itai-itai* disease is caused by environmental cadmium exposure, but the pathological mechanism of the disease remains unknown. Cadmium also yields irreversible bone and renal dysfunction; new validated biomarkers are needed for the detection of cadmium-induced nephropathy. We have focused on the role of fibroblast growth factor (FGF) 23, a protein that is essential for phosphate homeostasis in the bone-kidney axis, and have investigated the mechanism of cadmium-induced FGF23 production by bone cells. Cadmium injection in mice resulted in increased plasma FGF23 concentrations, but the level of *FGF23* mRNA in the bone was not changed. Further studies indicated that increased plasma FGF23 levels in the cadmium-injected mice were caused by the posttranslational regulation of the FGF23 protein stability. FGF23 stability and secretion was altered by glycosylation of FGF23, which was in turn regulated by the activity of the GalNAc-T3 protein. We demonstrated that expression of the GalNAc-T3-encoding gene was significantly increased by cadmium exposure. Moreover, cadmium-dependent FGF23 accumulation was inhibited by an antagonist of the aryl hydrocarbon receptor (AhR), a transcription factor that can bind to the promoter of the *GalNAc-T3* gene. Thus, cadmium stimulates transcription of the *GalNAc-T3* gene via enhanced binding of AhR to the *GalNAc-T3* promoter.

These findings suggest that the cadmium-induced increase in GalNAc-T3 expression suppresses proteolytic intracellular processing of FGF23, thereby yielding

S. Kido (✉)
Laboratory of Clinical Nutrition, Department of Food Science and Nutrition, Faculty of Agriculture, Kindai University, Nara, Japan

Agricultural Technology and Innovation Research Institute, Kindai University, Nara, Japan
e-mail: skido@nara.kindai.ac.jp

I. Kaneko · K.-i. Miyamoto
Department of Molecular Nutrition, Institute of Biomedical Sciences, Tokushima University Graduate School, Tokushima, Japan

S. Himeno, K. Aoshima (eds.), *Cadmium Toxicity*, Current Topics in Environmental Health and Preventive Medicine,
https://doi.org/10.1007/978-981-13-3630-0_14

increased serum FGF23 concentrations, which in turn alter phosphate metabolism and dysfunction of the kidney and bone.

Keywords Cadmium · Fibroblast growth factor 23 · Processing · Phosphaturia Bone

14.1 Inorganic Phosphate Homeostasis and Phosphate Transporters (NaPi)

Inorganic phosphate (Pi) is an essential nutrient that is required for normal skeletal mineralization and various cellular metabolic processes. In human and rodents, Pi homeostasis is controlled mainly by the action of three major organs (kidney, intestine, and bone). Among these organs, the kidney is the major regulator of Pi homeostasis; the sodium-dependent phosphate (NaPi) cotransporters found in the brush border membrane (BBM) are central to the reabsorption of Pi in renal proximal tubules [1, 2]. The solute carriers of three different types of renal sodium-dependent phosphate cotransporters—known as SLC34A1 (NaPi-IIa/NPT2a), SLC34A3 (NaPi-IIc/NPT2c), and SLC20A2 (PiT-2)—are specifically expressed on the BBM of the renal proximal tubules in rat and mouse [3, 4]. Among these cotransporters, Npt2a plays a major role in renal Pi reabsorption, while Npt2c is important for Pi reabsorption in weaning animals [5, 6]. The third NaPi transporter group, PiT-2, has been shown to be expressed in the proximal tubular cells [7]. Multiple factors regulate the function of these transporters, and an elaborate network of humoral interactions and feedback loops involving multiple organs (including intestine, kidney, and bones) is well-known.

14.2 Regulation of Phosphate Metabolism by FGF23

Some investigators argue that the bone is an endocrine gland, as evidenced by the fact that fibroblast growth factor 23 (FGF23) is produced by the bone, especially by osteocytes [8, 9]. FGF23 was originally identified as a phosphaturic factor that is expressed primarily in the bone [10, 11]. Physiologically, FGF23 is known to facilitate renal Pi excretion by suppressing Pi reabsorption within the renal proximal tubules; FGF23 also decreases plasma 1,25-dihydroxyvitamin D_3 levels by decreasing de novo synthesis of vitamin D [12]. The recently identified FGF23 co-receptor α-Klotho (α-KL) is necessary for FGF23 to activate FGF-specific receptors (FGFRs) and thereby induce downstream signaling [13, 14]; this FGF23-Klotho pathway is essential for maintaining phosphate and vitamin D metabolism [15, 16]. There is clear evidence from numerous diseases and disease models that FGF23 plays a critical role in physiological phosphate metabolism [17]. Within the body, FGF23 acts as a circulating phosphaturic factor that controls systemic phosphate homeostasis by regulating renal Pi excretion [10, 11]. Mechanistically,

FGF23 decreases serum phosphate concentrations by suppressing the expression of NPT2a and NPT2c in the renal proximal tubules while also suppressing intestinal phosphate absorption by decreasing 1,25-dihydroxyvitamin D_3 levels by reducing the expression of rate-limiting enzymes in the vitamin D synthesis pathway [18]. Therefore, FGF23 is a potent systemic regulator of Pi and vitamin D homeostasis.

14.3 FGF23 and Abnormal Phosphate Metabolism in CKD

Chronic kidney disease (CKD) is the progressive and irreversible damaging of the kidneys over a period of years. Each human kidney contains about 1 million of the tiny filtering units called nephrons. Any disease that injures the nephrons can cause CKD; notably, high blood pressure and diabetes are both known to damage nephrons. Once lost, kidney function cannot be regained; thus, early detection and treatment of kidney damage and CKD are critically important.

Recently, FGF23 was identified as a potential regulator of the development of abnormal mineral and bone metabolism and associated comorbidities in CKD patients. Furthermore, various studies indicate that serum FGF23 levels are elevated early in kidney disease and are independently associated with increased morbidity and mortality [19–22]. Based on these results, FGF23 is expected to serve as an effective biomarker for early detection of CKD [23]. Additionally, inflammation is among the pathological conditions in CKD patients and has emerged as a novel regulator of FGF23 levels. Notably, recent reports have shown that inflammation leads to increased FGF23 production and may contribute to elevated FGF23 levels in CKD [24, 25].

14.4 Cadmium Absorption and Toxicity in the Human Body

Cadmium is a major nephrotoxic heavy metal; this element is located in the periodic table between zinc and mercury and exhibits chemical properties similar to those of zinc. Cadmium is used industrially in the plating of steel, in pigments used in plastics and glasses, as an electrode material in nickel-cadmium batteries, and as a component of various alloys [26–28]. Cadmium is found as an impurity in zinc and lead deposits and therefore is produced primarily as a by-product of zinc or lead smelting. As a result, cadmium contamination is often a serious problem around mines that produce zinc and lead. Cadmium exposure occurs from ingestion of cadmium-contaminated food (e.g., organ meats, leafy vegetables, fruits, and seeds) or water, including groundwater and rivers flowing near the contaminated area. Cigarette smoking is also considered to be a significant source of cadmium pollution [29]. When cadmium is taken into the body, cadmium is transported throughout the body via binding to metallothionein, a sulfhydryl group-containing protein [30, 31].

Much of the incorporated cadmium is located in the liver and kidney, with the balance distributed throughout the body. Since the biologic half-life of cadmium is quite long (more than 30 years), prolonged low-level exposure leads to excessive accumulation in certain tissues, especially the kidney [32]. The toxicity of cadmium in the body is very diverse, with effects including oxidative stress [33], epigenetic changes in DNA expression [34, 35], and serious damage to various transport pathways, particularly in the proximal S1 segment of the renal proximal tubules, a condition called Fanconi syndrome [36]. Fanconi syndrome is a disorder of the renal proximal tubules in which certain substances normally absorbed into the bloodstream by the kidneys are released into the urine. In addition, clinical laboratory tests of patients with Fanconi syndrome may reveal excess urinary excretion of various substances, including amino acids, bicarbonate, glucose, magnesium, potassium, sodium, and phosphate.

14.5 *Itai-itai* Disease: Causes, Symptoms, and Clinical Features

Itai-itai disease was one of the "big four" pollution diseases of Japan (the other three being Minamata disease, Niigata Minamata disease, and Yokkaichi asthma). The name of the *itai-itai* disease (literally, "it hurts, it hurts" in Japanese) is imitative of the sufferers' screams due to intolerable bone pain. Historically, *itai-itai* was observed primarily in the Jinzu River basin region in the Toyama Prefecture, where the incidence was noted starting in the 1930s [37]. The disease was designated as *itai-itai* in 1958 and was finally acknowledged as the first pollution disease in 1968. However, initially, the cause of the disease was unknown, and the condition was considered to be an endemic disease.

The Jinzu River basin had been the site of mining and refining of zinc (and other metals) for centuries; the resulting mining waste and slag (which included cadmium) were dumped directly into the Jinzu River. As a result, cadmium and other heavy metals accumulated in the river sediments. Water from the river was used primarily for irrigation of rice fields; downstream inhabitants also used the river as a source of drinking water, fish, and washing [38]. The rice grown in this water incorporated various heavy metal pollutants, especially cadmium. Ultimately, many of the downstream inhabitants who consumed this contaminated rice, water, and fish developed *itai-itai* disease.

The main clinical features of *itai-itai* disease were osteomalacia accompanied by osteoporosis and impairment of multiple proximal tubular functions; the disease was considered to be a type of acquired Fanconi syndrome that resulted in markedly increased excretion of urinary protein, glucose, and amino acids by patients. Interestingly, a marked prevalence in older, postmenopausal women was observed [39, 40]. However, the main cause of this particular phenomenon remains poorly understood and is currently the subject of ongoing discussion.

Since 1967, 184 victims have been legally recognized, of whom 54 were recognized in the period from 1980 to 2000. However, an additional group of approximately 400 people have been identified as potential victims, with this population corresponding to individuals who have not yet been officially examined. On the other hand, although cadmium has been removed from the polluted water in Japan, this metal is now being used in many new industrial products, not only in Japan but all over the world. For example, some researchers theorize that widespread cadmium pollution in Hunan Province, South China, poses an ongoing threat to public health.

14.6 Cadmium Exposure and FGF23

Given that renal tubular dysfunction associated with cadmium is irreversible [41], more sensitive and specific biomarkers are urgently required; existing markers (e.g., proteinuria) have limitations as biomarkers of cadmium-induced nephropathy and responses to interventions. Therefore, new validated biomarkers of cadmium-induced nephropathy are needed. We have focused on FGF23, an essential component of the phosphate homeostasis pathway in the bone-kidney axis, as a potential biomarker. Specifically, we investigated whether FGF23 is involved in cadmium-induced nephropathy using cadmium-injected mice as an animal model.

Following cadmium injection in mice, the accumulation of cadmium in various tissues was measured using flame atomic absorption spectrometry. Markedly high concentrations of cadmium were observed in the liver and kidney, consistent with a previous report [42]. Cadmium also accumulated in the bone, the known FGF23-producing organ, but the levels in the bone were slight compared to those in the kidney and liver [43].

Mice injected with cadmium showed decreases in body weight compared to control animals, even though food intake and urine volume differed little between the two groups. The cadmium-injected mice also showed significant increases in plasma levels of calcium, blood urea nitrogen (BUN), plasma creatinine (Cr), and plasma glucose, along with a slight decrease in plasma phosphate concentration. In contrast, no significant changes were observed in the concentrations of plasma parathyroid hormone or 1,25-dihydroxyvitamin D_3 upon cadmium injection. Interestingly, plasma concentrations of FGF23 were significantly increased in the cadmium-injected mice (compared to control animals). Taken together, these results suggested that although cadmium injection in mice causes only mild renal dysfunction, FGF23 is involved in acute or subacute cadmium toxicity in the kidney and bone [43]. This work represents the first report (to our knowledge) of a relationship between cadmium toxicity and FGF23 levels.

We next examined the effect of cadmium on NaPi cotransporters. Specifically, NaPi transport activity was evaluated by measuring the expression level of NaPi in BBM vesicles isolated from vehicle- or cadmium-injected mice. We found that NaPi cotransporter activity was significantly decreased in the kidneys of cadmium-treated mice. To clarify the molecular mechanisms underlying the cadmium-induced decrease in phosphate transport activity, we analyzed the expression of Npt2a and Npt2c, which are the major phosphate transporters expressed in the renal proximal tubule. Notably, decreased phosphate transport was associated with a decrease in expression (at both the protein and mRNA levels) of the renal phosphate transporter Npt2c, but not that of Npt2a [43]. These observations suggested that the renal toxicity of cadmium exposure is mediated via suppression of the expression of phosphate transporters, especially that of Npt2c, in the renal proximal tubule, resulting in decreased phosphate transport activity.

The effect of cadmium on phosphate metabolism has been reported previously [44], and FGF23 is known to cause decreased serum phosphate levels by suppressing the expression of Npt2a and Npt2c in the renal proximal tubules [45]. Based on these results, FGF23 is expected to mediate cadmium-induced hypophosphatemia, but the detailed mechanism of this effect remained unknown.

14.7 Regulation of FGF23 Expression in Bone Cells by Cadmium

We therefore examined the effect of cadmium exposure on *FGF23* gene expression and plasma FGF23 concentrations in vivo. We found that plasma FGF23 concentrations were significantly increased following administration of cadmium, but the levels of *FGF23* mRNA in the bone were not elevated [43]. More recently, we confirmed that cadmium does not induce the expression of *FGF23* mRNA under either in vitro or in vivo conditions [46]. These observations suggested that the increase in FGF23 production upon cadmium exposure involves posttranscriptional regulation.

Regarding the mechanism of increased FGF23 production by the bone, bone-synthesized FGF23 protein is known to be cleaved between Arg^{179} and Ser^{180} during the process of secretion, resulting in inactive N- and C-terminal fragments [11]. Moreover, White et al. demonstrated that mutations that alter the proteolytic cleavage sites of FGF23 (R176Q/W and R179Q/W) prevent processing and result in autosomal dominant hypophosphatemic rickets (ADHR) [10]. It is evident from these data that FGF23 activity is controlled not only by de novo synthesis but also by posttranslational modification of the FGF23 protein. These observations suggested that changes in FGF23 processing might be important for regulating plasma FGF23 levels.

14.8 Cadmium Reduces FGF23 Processing via GalNAc-T3 in the Bone

GalNAc-T3 is a member of the UDP-N-acetyl-α-D-galactosamine/polypeptide N-acetylgalactosaminyltransferase (GalNAc-T) family, GalNAc-T3 *O*-glycosylate FGF23 protein at a furin-like convertase recognition sequence, and we hypothesized that the modified FGF23 is resistant to the cleavage described above [47, 48].

In brief, induction of transcription of the GalNAc-T3-encoding gene by cadmium results in an increase in the circulating level of intact FGF23, leading in turn to urinary phosphate excretion. To test this hypothesis directly, we investigated whether cadmium exposure induced *GalNAc-T3* mRNA expression in the bone in cadmium-injected mice. We found that cadmium exposure enhances the mRNA expression of the *GalNAc-T3* gene in the bone in vivo. This observation was consistent with previous reports in which GalNAc-T3 and FGF23 were found to be co-expressed in several tissues [49]. Moreover, a previous report revealed that FGF23 is expressed predominantly in osteocytes in the bone [9]. We have confirmed that cadmium-induced *GalNAc-T3* gene expression is observed in osteocytes by testing primary osteocytes isolated from the calvariae of newborn mice [46]. Moreover, immunohistochemical analysis demonstrated that both FGF23 and GalNAc-T3 are expressed in both trabecular and cortical bone osteocytes, as shown in a previous report [49]. The induction of FGF23 production is mediated via the increased accumulation of GalNAc-T3 following cadmium exposure; knockdown of GalNAc-T3 production by siRNA enhances FGF23 processing in bone cells [46]. These results are concordant with our hypothesis that cadmium exposure promotes the stabilization of FGF23 through the induction of *GalNAc-T3* gene expression in bone cells.

14.9 Cadmium Induces Increased FGF23 Production by Inducing *GalNAc-T3* Gene Transcription via AhR in Bone Cells

The aryl hydrocarbon receptor (AhR) is a ligand-activated helix-loop-helix transcription factor that binds various exogenous ligands, including natural plant flavonoids, polyphenolics, and indoles, as well as synthetic polycyclic aromatic hydrocarbons (PAHs) and dioxin-like compounds. AhR is a cytosolic protein that is normally inactive and bound to several co-chaperones. However, upon ligand binding (e.g., to chemicals such as TCDD (2,3,7,8-tetrachlorodibenzo-*p*-dioxin), the so-called dioxin), the chaperones dissociate from AhR. AhR then translocates into the nucleus and dimerizes with the AhR nuclear translocator (ARNT); the resulting complex binds to DNA promoter sequences termed xenobiotic response elements (XREs), thereby enhancing the transcription of target genes [50]. As noted above, AhR ligands include hazardous environmental pollutants such as dioxin and its analogues. Previous reports demonstrated that several heavy metals, including cadmium, can induce AhR activation, thereby facilitating increased expression of target genes, including the locus encoding cytochrome P450 1A1 (CYP1A1) [51, 52]. Based on this mechanism, various heavy metals are considered endocrine-disrupting chemicals [53, 54].

Given that AhR plays a central role in endocrine disruption, we investigated whether AhR is involved in the cadmium-based induction of GalNAc-T3 in bone cells. First, we examined the effect of a selective AhR antagonist, 3,4-dimethoxyflavone (DMF), on *GalNAc-T3* induction by cadmium exposure. Induction of *GalNAc-T3* mRNA by cadmium was almost completely abrogated by

3,4-DMF in a dose-dependent manner. Moreover, knockdown of endogenous AhR by siRNA almost completely abrogated the induction of *GalNAc-T3* mRNA expression by cadmium in bone cells. These observations suggested that cadmium's induction of *GalNAc-T3* gene expression in bone cells is mediated via activation of AhR. In addition, we found that endogenous AhR was translocated into the nucleus by cadmium exposure and that both activation and nuclear localization are mediated by the p38-MAPK signaling pathway [46].

To further investigate the role of AhR, we examined the promoter sequences of the human *GalNAc-T3* gene [55]. We confirmed that the increase in *GalNAc-T3* gene transcription is mediated via the p38-MAPK-AhR signaling pathway, using a reporter-gene assay to show that activated AhR stimulates *GalNAc-T3* gene transcription through binding to a putative XRE-like site (5′-TGGCGCG-3′) found in the *GalNAc-T3* gene promoter [46]. Studies with site-directed mutagenesis at the putative XRE sites revealed that XRE is required for activation of *GalNAc-T3* gene promoter activity by cadmium. These observations demonstrated that p38-MAPK induction of AhR activation is necessary for stimulation of *GalNAc-T3* gene transcription in bone cells following cadmium exposure.

14.10 Conclusion

In this chapter, we described the identification of a new signaling pathway that permits bone cells to respond to cadmium exposure. Exposure to high levels of cadmium enhances *GalNAc-T3* gene expression, which leads to inhibition of FGF23 cleavage and permits the accumulation of intact FGF23 in bone cells. Our data further suggested that altered FGF23 processing is involved in the cadmium-dependent increase in FGF23 production in bone cells. Previous reports demonstrated that various external stimuli induce *GalNAc-T3* gene expression, including Pi, calcium, and vitamin D [56], but the molecular mechanisms of these induction events are not fully understood [57]. In the present study, we demonstrated that *GalNAc-T3* gene expression was altered by cadmium exposure, both in vitro (in bone cells) and in vivo (in mice). Signal transduction from cadmium exposure involves increased intracellular p38-MAPK signaling pathway, which activates AhR and facilitates nuclear translocation; AhR then binds to the XRE-like sites in the *GalNAc-T3* gene promoter to enhance transcription of this locus. Activation of *GalNAc-T3* gene transcription causes increased of GalNAc-T3 protein levels; the resulting glycosylation of FGF23 inhibits cleavage of that protein and thereby yields increased FGF23 levels in bone cells, as shown in Fig. 14.1. Moreover, elevation of FGF23 concentration also inhibits renal Pi reabsorption and vitamin D synthesis in the renal proximal tubular cells [58].

In conclusion, the work summarized here demonstrates that the p38-MAPK~AhR pathway plays a critical role in intracellular signaling in response to cadmium exposure. These results additionally clarify the role of FGF23 in the pathogenesis of *itai-itai* disease.

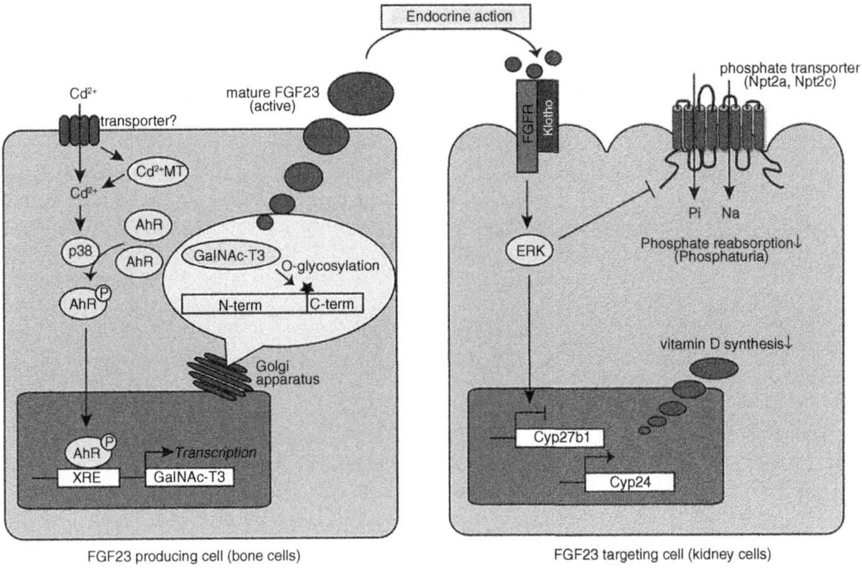

Fig. 14.1 Mechanism of cadmium-induced FGF23 production in bone cells. Cadmium activates and stimulates nuclear translocation via p38-MAPK, thereby enhancing *GalNAc-T3* gene transcription. The increased GalNAc-T3 protein levels lead to glycosylation of FGF23, suppressing proteolytic processing of FGF23 and resulting in the elevation of blood FGF23 concentrations. FGF23 in turn suppresses renal NaPi-II transporter and vitamin D synthesis in the kidney, effects that may cause phosphaturia along with renal and bone dysfunction

References

1. Murer H, Silve C, Biber J. Phosphate metabolism renal handling of phosphate. In: Massry & glassock textbook of nephrology. 4th ed. Philadelphia: Lippincott Williams & Wilkins.
2. Murer H, Hernando N, Forster I, Biber J. Proximal tubular phosphate reabsorption: molecular mechanisms. Physiol Rev. 2000;80:1373–409.
3. Tenenhouse HS. Regulation of phosphorus homeostasis by the type IIa Na/phosphate cotransporter. Annu Rev Nutr. 2005;25:197–214.
4. Biber J, Hernando N, Forster I, Murer H. Regulation of phosphate transport in proximal tubules. Pflugers Arch. 2009;458:39–52.
5. Segawa H, Kaneko I, Takahashi A, Kuwahata M, Ito M, Ohkido I, Tatsumi S, Miyamoto K. Growth-related renal type II Na/pi cotransporter. J Biol Chem. 2002;277:19665–72.
6. Miyamoto K, Ito M, Tatsumi S, Kuwahata M, Segawa H. New aspect of renal phosphate reabsorption: the type II c sodium-dependent phosphate transporter. Am J Nephrol. 2007;27:503–15.
7. Breusegem SY, Takahashi H, Girl-Arnal H, Wang X, Jiang T, Verlander JW, Wilson P, Miyazaki-Anzai S, Sutherland E, Caldas Y, Blaine JT, Segawa H, Miyamoto K, Barry NP, Levi M. Differential regulation of the renal sodium-phosphate cotransporters NaPi-IIa, NaPi-IIc, and PiT-2 in dietary potassium deficiency. Am J Physiol Renal Physiol. 2009;297:F350–61.
8. Feng JQ, Ward LM, Liu S, Lu Y, Xie Y, Yuan B, Yu X, Rauch F, Davis SI, Zhang S, Rios H, Drezner MK, Quarles LD, Bonewald LF, White KE. Loss of DMP1 causes rickets and osteomalacia and identifies a role for osteocytes in mineral metabolism. Nat Genet. 2006;38:1310–5.
9. Liu S, Zhou J, Tang W, Jiang X, Rowe DW, Quarles LD. Pathogenic role of fgf23 in hyp mice. Am J Physiol Endocrinol Metab. 2006;291:E38–49.

10. White KE, Evans WE, O'Riordan J, Speer MC, Econs MJ, Lorenz-Depiereux B, Grabowski M, Meitinger TTMS. Autosomal dominant hypophosphatemic rickets is associated with mutations in FGF23. Nat Genet. 2000;26:345–8.
11. Shimada T, Mizutani S, Muto T, Yoneya T, Hino R, Takeda S, Fujita T, Fukumoto S, Yamashita T. Cloning and characterization of FGF23 as a causative factor of tumor-induced osteomalacia. Proc Natl Acad Sci U S A. 2001;98:6500–5.
12. Inoue Y, Segawa H, Kaneko I, Yamanaka S, Kusano K, Kawakami E, Furutani J, Ito M, Kuwahara M, Saito H, Fukushima N, Kato S, Kanayama HO, Miyamoto K. Role of the vitamin D receptor in FGF23 action on phosphate metabolism. Biochem J. 2005;390:325–31.
13. Kurosu H, Ogawa Y, Miyoshi M, Yamamoto M, Nandi A, Rosenblatt KP, Baum MG, Schiavi S, Hu MC, Moe OW, Kuro-o M. Regulation of fibroblast growth factor-23 signaling by klotho. J Biol Chem. 2006;281:6120–3.
14. Urakawa I, Yamazaki Y, Shimada T, Iijima K, Hasegawa H, Okawa K, Fujita T, Fukumoto S, Yamashita T. Klotho converts canonical FGF receptor into a specific receptor for FGF23. Nature. 2006;444:770–4.
15. Murer H, Forster I, Biber J. The sodium phosphate cotransporter SLC34. Pflugers Arch. 2004;447:763–7.
16. Tatsumi S, Miyagawa A, Kaneko I, Shiozaki Y, Sagawa H, Miyamoto K. Regulation of renal phosphate handling: inter-organ communication in health and disease. J Bone Miner Metab. 2016;34:1–10.
17. Erben RG. Update on FGF23 and klotho signaling. Mol Cell Endocrinol. 2016;432:56–65.
18. Shimada T, Hasegawa H, Yamazaki Y, Muto T, Hino R, Takeuchi Y, Fujita T, Nakahara K, Fukumoto S, Yamashita T. FGF-23 is a potent regulator of vitamin D metabolism and phosphate homeostasis. J Bone Miner Res. 2003;19:429–35.
19. Guiterrez OM, Januzzi JL, Isakova T, Laliberte K, Smith K, Collerone G, Sarwar A, Hoffmann U, Coglianese E, Christensen R, Wang TJ, deFilippi C, Wolf M. Fibroblast growth factor 23 and left ventricular hypertrophy in chronic kidney disease. Circulation. 2009;119:2545–52.
20. Gutierrez OM, Mannsrtadt M, Isakova T, Rauh-Hain JA, Tamez H, Shah A, Smith K, Lee H, Thadhani R, Juppner H, Wolf M. Fibroblast growth factor 23 and mortality among patients undergoing hemodialysis. N Engl J Med. 2008;359:584–92.
21. Mirza MA, Larsson A, Melhus H, Ling L, Lasson TE. Serum intact FGF23 associate with left ventricular mass, hypertrophy and geometry in an elderly population. Atherosclerosis. 2009;207:546–51.
22. Gutierrez O, Isakova T, Rhee E, Shah A, Holmes J, Collerone G, Juppner H, Wolf M. Fibroblast growth factor 23 mitigates hyperphosphatemia but accentuates calcitriol deficiency in chronic kidney disease. J Am Soc Nephrol. 2005;16:2205–15.
23. Wahl P, Wolf M. FGF23 in chronic kidney disease. Adv Exp Med Biol. 2012;728:107–25.
24. Munoz Mendoza J, Isakova T, Ricardo AC, Xie H, Navaneethan SD, Anderson AH, Bezzano LA, Xie D, Kretzler M, Nessel L, Hamm LL, Negrea L, Leonard MB, Raj D, Wolf M, Chronic Renal Insufficiency Cohort. Fibroblast growth factor 23 and inflammation in CKD. Clin J Am Soc Nephrol. 2012;7:1155–62.
25. Francis C, David V. Inflammations regulates fibroblast growth factor 23 production. Curr Opin Nephrol Hypertens. 2016;25:325–32.
26. Adams RG, Harrison JF, Scott P. The development of cadmium-induced proteinuria, impaired renal function, and osteomalacia in alkaline battery workers. Q J Med. 1969;38:425–43.
27. Kazantzis G, Flynn FV, Spowage JS, Trott DG. Renal tubular malfunction and pulmonary emphysema in cadmium pigment workers. Q J Med. 1963;32:165–92.
28. Friberg L. Cadmium and the kidney. Environ Health Perspect. 1984;54:1–11.
29. Friberg L, Piscator M, Nordberg GF, Kjellstrom T. Cadmium. Annu Rev Public Health. 1983;4:367–73.
30. Klaassen CD, Liu J. Metallothionein transgenic and knock-out mouse models in the study of cadmium toxicity. J Toxicol Sci. 1998;23:97–102.

31. Klaassen CD, Liu J, Diwan BA. Metallothionein protection of cadmium toxicity. Toxicol Appl Pharmacol. 2009;238:215–20.
32. Gonick HC. Trace metals and the kidney. Miner Electrolyte Metab. 1978;1:107–20.
33. Cuypers A, Plusquin M, Remans T, Jozefczak M, Keunen E, Gielen H, Qpdenakker K, Nair AR, Munters E, Artois TJ, Nawrot T, Vangronsveld J, Smeets K. Cadmium stress: an oxidative challenge. Biometals. 2010;23:927–40.
34. Wang B, Li Y, Shao C, Tan Y, Cai L. Cadmium and its epigenetic effects. Curr Med Chem. 2012;19:2611–20.
35. Martinez-Zamudio R, Ha HC. Environmental epigenetic in metal exposure. Epigenetics. 2011;6:820–7.
36. Vesey DA. Transport pathways for cadmium in the intestine and kidney proximal tubule: focus on the interaction with essential metals. Toxicol Lett. 2010;198:13–9.
37. Almeida P, Stearns L. Political opportunities and local grassroots environmental movement: the case of Minamata. Soc Probl. 1998;45:37–60.
38. Kaneta M, Hikichi H, Endo S, Sugiyama N. Chemical form of cadmium (and other heavy metals) in rice and wheat plants. Environ Health Persp. 1986;65:33–7.
39. Johri N, Jacquillet G, Unwin R. Heavy metals poisoning: the effects of cadmium on the kidney. Biometals. 2010;23:783–92.
40. Kazantzis G. Cadmium, osteoporosis and calcium metabolism. Biometals. 2004;17:493–8.
41. Iwata K, Saito H, Moriyama M, Nakano A. Renal tubular function after reduction of environmental cadmium exposure: a ten year follow up. Arch Environ Health. 1993;48:157–63.
42. Lind Y, Engman J, Jorhem L, Glynn AW. Cadmium accumulation in liver and kidney of mice exposed to the same weekly cadmium dose continuously or once a week. Food Chem Toxicol. 1997;35:891–5.
43. Aranami F, Segawa H, Furutani J, Kuwahara S, Tominaga R, Hanabusa E, Tatsumi S, Kido S, Ito M, Miyamoto K. Fibroblast growth factor 23 mediates the phosphaturic actions of cadmium. J Med Investig. 2010;57:95–108.
44. Herak-Kramberger CM, Spindler B, Biber J, Murer H, Sabolic I. Renal type II Na/pi co-transporter is strongly impaired whereas the Na/sulphate-cotransporter and aquaporin I are unchanged in cadmium-treated rats. Pflugers Arch. 1996;432:336–44.
45. Olauson H, Vervloet MG, Cozzolino M, Massey ZA, Urena Torres P, Larsson TE. New insights into the FGF23-Klotho axis. Semin Nephrol. 2014;34:586–97.
46. Kido S, Fujihara M, Nomura K, Sasaki S, Mukai R, Ohnishi R, Kaneko I, Segawa H, Tatsumi S, Izumi H, Kohno K, Miyamoto K. Molecular mechanisms of cadmium-induced fibroblast growth factor 23 upregulation in osteoblast-like cells. Toxicol Sci. 2014;139:301–16.
47. Frishberg Y, Ito N, Rinat C, Yamazaki Y, feinsteins S, Urakawa I, Navon-Elkan P, Becker-Cohen R, Yamashita T, Araya K, Igarashi T, Fujita T, Fukumoto S. Hyperostosis-hyperphosphatemia syndrome: a congenital disorder of *O*-glycosylation associated with augment processing of fibroblast growth factor 23. J Bone Miner Res. 2007,22.235–42.
48. Kato K, Jeanneau C, Tarp MA, Benet-Pages A, Lorenz-Depiereux B, Bennett EP, Mandel U, Strom TM, Clausen H. Polypeptide GalNAc-transferase T3 and familial tumoral calcinosis. Secretion of fibroblast growth factor 23 requires O-glycosylation. J Biol Chem. 2006;281:18370–7.
49. Ichikawa S, Lyles KW, Econs MJ. A novel GALNT3 mutation in a pseudoautosomal dominant form of tumoral calcinosis: evidence that the disorder is autosomal recessive. J Clin Endocrinol Metab. 2005;90:2420–3.
50. Carlson DB, Perdew GH. A dynamic role for the Ah receptor in cell signaling? Insights from a diverse group of Ah receptor interacting proteins. J Biochem. 2002;16:317–25.
51. Elbekai RH, El-Kadi AO. Modulation of aryl hydrocarbon receptor-regulated gene expression by arsenite, cadmium, and chromium. Toxicology. 2004;202:249–69.
52. Elbekai RH, El-Kadi AO. Transcriptional activation and posttranscriptional modification of cyp1a1 by arsenite, cadmium, and chromium. Toxicol Lett. 2007;172:106–19.

53. Johnson MD, Kenny N, Stoica A, Hilakivi-Clarke L, Singh B, Chepko G, Clarke R, Sholler PF, Lirio AA, Foss C, Reiter R, Trock B, Paik S, Martin MB. Cadmium mimics the *in vivo* effects of estrogen in the uterus and mammary gland. Nat Med. 2003;9:1081–4.
54. Martines-Campa C, Alonso-Gonzalez C, Mediavilla MD, Cos S, Gonzalez A, Ramos S, Sanchez-Barcelo EJ. Melatonin inhibits both ER alpha activation and breast cancer cell proliferation induced by a metalloestrogen, cadmium. J Pineal Res. 2006;40:291–6.
55. Nomoto M, Izumi H, Ise T, Kato K, Takano H, Nagatani G, Shibao K, Ohta R, Imamura T, Kuwano M, Matsuo K, Yamada Y, Itoh H, Kohno K. Structural basis for the regulation of UDP-N-acetyl-alpha-D-galactosamine: polypeptide N-acetylgalactosaminyl transferase-3 gene expression in adenocarcinoma cells. Cancer Res. 1999;59:6214–22.
56. Chefetz I, Kohno K, Izumi H, Uitto J, Richard G, Sprecher E. GALNT3, a gene associated with hyperphosphatemic familial tumoral calcinosis, is transcriptionally regulated by extracellular phosphate and modulates matrix metalloproteinase activity. Biochem Biophys Acta. 2009;1792:61–7.
57. Chefetz I, Sprecher E. Familial tumoral calcinosis and the role of O-glycosylation in the maintenance of phosphate homeostasis. Biochem Biophys Acta. 2009;1792:847–52.
58. Kido S, Kaneko I, Tatsumi S, Segawa H, Miyamoto K. Vitamin D and type II sodium-dependent phosphate cotransporters. Contrib Nephrol. 2013;180:86–97.

Correction to: Recent Clinical and Epidemiological Studies of *Itai-Itai* Disease (Cadmium-Induced Renal Tubular Osteomalacia) and Cadmium Nephropathy in the Jinzu River Basin in Toyama Prefecture, Japan

Keiko Aoshima

Correction to:
Chapter 2 in: S. Himeno, K. Aoshima (eds.), *Cadmium Toxicity*, Current Topics in Environmental Health and Preventive Medicine, https://doi.org/10.1007/978-981-13-3630-0

This chapter was inadvertently published with incorrect data in the abstract. Details are as follows:
Incorrect: < 1 mg/g creatinine
Correct: < 1%

The updated online version of this chapter can be found at
https://doi.org/10.1007/978-981-13-3630-0_2

S. Himeno, K. Aoshima (eds.), *Cadmium Toxicity*, Current Topics in Environmental Health and Preventive Medicine,
https://doi.org/10.1007/978-981-13-3630-0_15

The manufacturer's authorised representative in the EU is Springer Nature Customer Service Centre GmbH, Europaplatz 3, 69115 Heidelberg, Germany. If you have any concerns regarding our products, please contact ProductSafety@springernature.com

Printed and bound by CPI Group (UK) Ltd, Croydon, CR0 4YY
30/09/2026
02225471-0014